AN INSIGHTS INTO
INDUSTRIAL MICROBIOLOGY

THE EDITORS

Dr. Latika Bhatia is an assistant professor in department of Microbiology and Bioinformatics in Atal Bihari Vajpayee Vishwavidyalaya, Bilaspur, C.G., India. She has 22 years of teaching experience working in university, teaching graduates and post graduates of microbiology and biotechnology. She did her Ph.D. from ITM University, Gwalior, on Biofuel Technology. She has published more than 35 research articles in peer-reviewed journals of an international repute, 10 book chapters and 03 monographs, 05 books, with citation index more than 600. Dr. Bhatia has developed online certificate course on immunology, in collaboration with Commonwealth Educational Media Center for Asia (CEMCA), available on Web-portal-"e-atalgyansangum", on UGC's specified 4 quadrant approaches. Dr. Bhatia is convener of Bilaspur Chapter of Indian Science Congress Association.

Dr. Prakash Kumar Sarangi is a Scientist with specialization in Microbiology at the Central Agricultural University in Imphal, India. He has a Ph.D. degree in Microbial Biotechnology from the Ravenshaw University, India and MTech. degree from Department of Agriculture and Food Engineering, Indian Institute of Technology Kharagpur, India. Dr. Sarangi's research is focused on bioprocess engineering, biochemicals, biomaterials and sustainable development. His expertise is in bioconversion of biomass into biofuels, biochemicals and nutraceuticals, as well as environmental waste remediation. He has published more than 60 research articles in peer-reviewed journals and authored in more than 25 book chapters.

AN INSIGHTS INTO
INDUSTRIAL MICROBIOLOGY

– Editors –

Dr. Latika Bhatia

Dr. Prakash Kumar Sarangi

2024

Daya Publishing House®

A Division of

Astral International Pvt. Ltd.

New Delhi – 110 002

ISBN: 978-93-5461-997-7

Publisher's Note:

Every possible effort has been made to ensure that the information contained in this book is accurate at the time of going to press, and the publisher and author cannot accept responsibility for any errors or omissions, however caused. No responsibility for loss or damage occasioned to any person acting, or refraining from action, as a result of the material in this publication can be accepted by the editor, the publisher or the author. The Publisher is not associated with any product or vendor mentioned in the book. The contents of this work are intended to further general scientific research, understanding and discussion only. Readers should consult with a specialist where appropriate.

Every effort has been made to trace the owners of copyright material used in this book, if any. The author and the publisher will be grateful for any omission brought to their notice for Acknowledgment in the future editions of the book.

Published by : **Daya Publishing House®**
A Division of
Astral International Pvt. Ltd.
– ISO 9001:2015 Certified Company –
4736/23, Ansari Road, Darya Ganj,
New Delhi-110 002
Ph. 011-43549197, 8130496929
E-mail: info@astralint.com
Website: www.astralint.com

Printed at : **Replika Press Pvt. Ltd.**

PREFACE

In the ever-evolving landscape of industry, the pivotal role played by microorganisms cannot be overstated. From the controlled fermentations that yield high-value products to the manipulation of genetic codes for improved performance, the strides made in Industrial Microbiology have revolutionized the way we produce, manufacture, and sustain our world. This book stands as a testament to the remarkable advancements that shape our future. Advances in Industrial Microbiology is not a static compendium; it is a dynamic exploration of the frontiers where science, engineering, and innovation converge. It illuminates the relentless efforts of scientists, researchers, and pioneers who, armed with curiosity and perseverance, push the boundaries of what is possible. Together, they uncover the hidden potential of microorganisms and translate it into practical solutions that propel industries forward. Within these chapters, we delve into the depths of genetic engineering, synthetic biology, and bioprocessing, where microbes are transformed into customizable workhorses. We witness the breath-taking applications of Industrial Microbiology in fields such as pharmaceuticals, biofuels, green chemistry, and waste management. We uncover the untapped potential of microbiomes, exploring how these intricate ecosystems hold the keys to sustainable agriculture, environmental remediation, and human health. But this book is more than a scientific exposition; it is a clarion call to embrace the microbial revolution. As we navigate the complexities of a rapidly changing world, we must recognize that microorganisms are not merely tools but partners in our journey toward a sustainable future. They offer us new avenues for resource efficiency, ecological harmony, and human well-being. It is our responsibility to harness their power wisely, guided by ethical considerations and a commitment to the greater good. Dear reader, prepare to embark on an enthralling expedition through the cutting edge of Industrial Microbiology. Embrace the microbial revolution, where science and ingenuity collide to unlock transformative possibilities. Welcome to the extraordinary world of Advances in Industrial Microbiology.

Dr. Latika Bhatia

Dr. Prakash Kumar Sarangi

CONTENT

LIST OF CONTRIBUTORS

Akanksha Kulshreshtha
Department of Biological Sciences and Engineering
Netaji Subhas University of Technology
Delhi, India

Ajay Kumar Kataria
Department of Biological Sciences and Engineering
Netaji Subhas University of Technology
Delhi, India

Annesha Chakraborty
Microbial Engineering & Algal Biotechnology Laboratory
Department of Biosciences
JIS University Kolkata
81 Nilgunj Road, Agarpara
Kolkata 700109, India

Ankita Gole
Microbial Engineering & Algal Biotechnology Laboratory
Department of Biosciences
JIS University Kolkata
81 Nilgunj Road, Agarpara
Kolkata 700109, India

Ayan Samanta
Microbial Engineering & Algal Biotechnology Laboratory
Department of Biosciences
JIS University Kolkata
81 Nilgunj Road, Agarpara
Kolkata 700109, India

Akhilesh Kumar Singh
Department of Biotechnology, School of Life Sciences, Mahatma Gandhi Central University, Motihari, Bihar-845401, India.

Akanksha Tiwari
Department of Food Processing & Technology
Atal Bihari Vajpayee University
Bilaspur, Chhattisgarh, India

Dowluru SVGK Kaladhar
Department of Microbiology and Bioinformatics
Atal Bihari Vajpayee University
Bilaspur, Chhattisgarh, India

Dilip Kumar Sahu
Department of Microbiology & Bioinformatics
Atal Bihari Vajpayee University
Bilaspur, Chhattisgarh, India

Dipankar Ghosh
Microbial Engineering and Algal Biotechnology Laboratory
Department of Biotechnology,
JIS University, Kolkata, West Bengal, India

Dona Roy
Microbial Engineering & Algal Biotechnology Laboratory
Department of Biosciences
JIS University Kolkata
81 Nilgunj Road, Agarpara
Kolkata 700109, India

Dhananjay Shukla
Department of Biotechnology,
Guru Ghasidas University,
Bilaspur Chhattisgarh, India

Harit Jha
Assistant Professor
Department of Biotechnology
Guru GhasidasVishwavidyalaya
Bilaspur, Chhattisgarh-495009
India

Kamalendu De
Microbial Engineering & Algal Biotechnology Laboratory
Department of Biosciences
JIS University Kolkata
81 Nilgunj Road, Agarpara
Kolkata 700109, India

Latika Bhatia
Department of Microbiology and Bioinformatics
Atal Bihari Vajpayee University
Bilaspur, Chhattisgarh, India

Ng. Joykumar Singh
College of Agriculture
Central Agricultural University
Imphal, Manipur, India

O.N. Tiwari
Centre for Conservation and Utilization of Blue Green Algae
Division of Microbiology
ICAR- Indian Agricultural Research Institute, New Delhi, India

Prakash Kumar Sarangi
College of Agriculture
Central Agricultural University
Imphal, Manipur, India

Rajesh Kumar Srivastava
Department of Biotechnology, GIT, Gandhi Institute of Technology and Management,
Visakhapatnam, India

Rishav Kar
Department of Microbiology
Ramakrishna Mission Vivekananda Centenary College
Rahara, Khardaha, West Bengal 700118, India

Soumya Sasmal
Department of Biotechnology
Vishva-Bharti
Shantiniketon 731235

Shreyasmita Guha
Microbial Engineering & Algal Biotechnology Laboratory
Department of Biosciences
JIS University Kolkata
81 Nilgunj Road, Agarpara
Kolkata 700109, India

Supriyo Mal
Microbial Engineering & Algal Biotechnology Laboratory
Department of Biosciences
JIS University Kolkata
81 Nilgunj Road, Agarpara
Kolkata 700109, India

Soumitra Tiwari
Department of Food Processing & Technology
Atal Bihari Vajpayee University
Bilaspur, Chhattisgarh, India

Saurabh Saxena
Nextgen Invitro Diagnostics,
Gurugram, Haryana, India
Th. Anand Singh
College of Agriculture
Central Agricultural University
Imphal, Manipur, India

Tanushree Sarkar
Department of Biotechnology
Guru GhasidasVishwavidyalaya
Bilaspur, Chhattisgarh-495009 India

Tantravahi Srinivasan
Department of Botany
Indira Gandhi National Tribal
University
Amarkantak, Madhya Pradesh, India

Uttam Kumar Sahoo
Department of Forestry, Mizoram
University,
Aizawl-796004, India

Vinit Singh Baghel
Department of Biotechnology,
Guru Ghasidas University,
Bilaspur Chhattisgarh, India

Chapter-1

FERMENTATION MEDIA AND INDUSTRIAL STERILIZATION SYSTEMS

Rajesh Kumar Srivastava[1], Prakash Kumar Sarangi[2], Akhilesh Kumar Singh[3] and Uttam Kumar Sahoo[4]*

[1]Department of Biotechnology, GIT, Gandhi Institute of Technology and Management, Visakhapatnam, India

[2]College of Agriculture, Central Agricultural University, Imphal, Manipur, India.

[3]Department of Biotechnology, School of Life Sciences, Mahatma Gandhi Central University, Motihari, Bihar-845401, India.

[4]Department of Forestry, Mizoram University, Aizawl-796004, India

**Corresponding author Email: rajeshksrivastava73@yahoo.co.in;*

Abstract

Most of biological processes like fermentation are completed via cultivation of suitable microbial system that is specific and capable of producing the different type of metabolites (primary/ secondary). These processes require the specific media for cultivation of specific and over-producer microbial system such as bacterial, fungal and cyanobacterial system. Further, specific media requirement are needed for cultivation of animal and plants cell that are specified for some particular metabolites (could be secondary/ primary). All these media are needed to complete the different type of fermentation processes in liquid/ solid forms, depending on microbial culture preference to promote the enhance metabolites production. Normally fermentation media are used at laboratory, pilot and also industrial scale metabolites production, depending on needs and infrastructures. Normally laboratory level media utilization is done for gaining basic level information about metabolites synthesis pattern, microbial growth and substrates rates and also byproducts synthesis. But, we can use fermentation media in shake flasks (100 ml to 1.0 liter capacities) and also bioreactor levels (from 1 to 10 liters capacity). Other aspect, media sterilization is important tasks for pure culture cultivation, specialized for particular metabolites synthesis during fermentation process that can occur at batch, fed-batch and continuous modes and also metabolite/ microbial growth nature. Number of metabolites such as ethanol, butanol, methanol, biogas, organic acids, enzymes and single cell production synthesis is reported from application of fermentative media. This chapter explores about basic fermentative media, specialized and industrial sterilization approaches that could help in fermentation process at high efficiency and cost effective modes.

Keywords: *Media, Fermentation, Microbial cells, Sterilization, Industrial process, Metabolites*

1.1. Introduction

There are various types of cultivation media, reported in different modes of fermentation process and also needs of specific microbial cells cultivations that are used for specific type of metabolites biosynthesis (Shang et al., 2012). We are known about different nature of fermentative media that are solid/ liquid forms and are utilized for cultivation of different types of microbial/ other biological cells growth and these are specialized and overproducer of our own interest fermentative metabolites. We know about different natures of fermentative media, reported in complex, semi-complex and defined form, depending on need of medium composition for the growth of biological systems (Shang et al., 2012; Molen et al., 2013).

In this context, the use of chemically defined media is reported to gain more popularity in some commercial levels of fermentation process utilization that are needed for preparation of some biological products (Miao et al., 2016). But, traditionally/ commonly modes, these media are not still developed in frequent manner that can utilize at industrial processes and they do not exhibit favorable characteristic/ properties at large scale utilization via compared to traditional complex media (El-Elimat et al., 2012). Now numbers of reviews and its related papers have come for application, development and practical considerations that have highlighted on process economic of fermentative in chemical defined media utilization at the industrial environment (Miao et al., 2016; El-Elimat et al., 2012).

At large scale any types of metabolites, industrial fermentation processes are now more popular to utilize for bulk and fine chemicals or pharmaceuticals biosynthesis. And fermentation process can cover broad range of metabolites hosts (Berendsen et al., 2013). In fermentation process, fermentation fungi are used for production of organic acids including citric acid production via *Aspergillus niger* strain as good example (Fatima et al., 2019). And second case, penicillin as one good example of secondary metabolites and is an antibiotics. It was discovered in 1928 and now is produced by *Penicillium chrysogenum* and *P. rubens* strains at large scale via using fermentation processes (Berendsen et al., 2013; Fatima et al., 2019).

Others metabolite like insulin is produced by recombinant strain *Escherichia coli* and also using yeast strain *Saccharomyces cerevisiae* as recombinant protein via using fermentation in different years of this process development (Rezende et al., 2012). Another best example is bioethanol biosynthesis as first (via using food grains or direct organic matters and second generation biofuel (via using non-food material/ complex organic matters) that will depend on types of raw materials used in industrial fermentation processes. Now it is dependent on technological asset for reducing our future dependence on chemical and products synthesis from fossil fuels utilization (Rezende et al., 2012; Fatima et al., 2019). There are many reports that discussed the benefits of optimal media utilization in fermentation process and also help in high yield/ rate and productivity of metabolites. And it was found in suitable concentration of compositions in fermentation media via promoting the growth of Phoma species (such as *P. loveata*) as potential fungal species (pathogen of potatoes plants), used for production of bioherbicide (Rohlfs et al., 2011).

In optimal concentration of fermentative media compositions, maximum growth of this fungus is reported. First time, synthetic media (it is containing glucose, peptone and yeast extract) used for this fungal growth with variations in other carbon and nitrogen sources (i.e., food-grade sucrose uses instead of glucose, peptone uses instead of peptone and yeast extract) with optimization tasks (Guzmán-Chávez et al., 2018). Finally first time optimal synthetic media was designed with glucose and peptone (each 20 $g.L^{-1}$) and yeast extract (7.5 $g.L^{-1}$) with initial pH value around 6.0. And Industrial media at optimized composition was found to contain sucrose (20$g.L^{-1}$) and CSL (8%) at pH: 6 (Viggiano et al., 2017; Guzmán-Chávez et al., 2018). And these conditions of fermentative media reported for maximum quantity of this fungal biomass 22 and 33 $g.L^{-1}$ at original and modified/ complex media respectively with high quantity of bioherbicide. This reported paper concludes for best industrial media than synthetic media with cost effective nature that can help in cheap bioproducts synthesis from industrial fermentation process (Klaic et al., 2015).

This chapter will discuss the media composition impacts and bioprocess dependency that can improve yield/ titer of bio-products via using appropriate system. This chapter also focuses the impact of sterilization approach that can help to maintain the sterility during bioprocesses.

1.2. Media Composition

Nowadays metabolites production from microbial systems with most efficient way has created more attentions and there are many efforts such as using of statically models methods for fermentative media optimization, genetic components and metabolic pathways modification tools for microbial strain improvement and process parameters optimization are very common to apply for metabolites titer, productivity and yield enhancement. And numbers of fermentation processes are used in this context that can help in biological routes biochemical/ products synthesis (Osama et al., 2013). Researchers have not reached at same maturity levels what we use traditional chemical routes of bioproducts. But, recently using engineering tools such as mathematical model, process control algorithm and optimization techniques have showed more success efforts in improved and more efficient for microbial fermentation processes (Zhong et al., 2014).

These approaches perspective start with a description of some of the most important engineering issues / challenges within industrial level fermentation technology and it can help to solve these issues via understanding at basic level these challenges. Then it can provide the advantages on optimal fermentation processes (Singh et al., 2012; Zhong et al., 2014). Many researchers have explored the different approaches of production media optimization that helped to enhance the yield and titer of metabolites at maximum levels. Among many/ wide range of techniques/ approaches of media optimization, one approach is discussed as one-factor-at- a-time to modern statistical and mathematical techniques. In this context, artificial neural network (ANN) and genetic algorithm (GA) is explored as best approaches of media optimization (Singh et al., 2009).

As, we know that every technique comes with its merits and demerits and even containing more numbers of drawback of some these approach can provide best results in fermentation process. In recent period, people are using the combination of optimization techniques that can give desirable results (Singh et al., 2017). Fermentation media optimization techniques helped in enhancing the metabolites biosynthesis during fermentation processes and their results was done for comparative analysis in merit and demerits of these techniques with various conventional and modern optimization approaches that helped in media components and their concentrations with improved titer and yield in reduced fermentation periods (Singh et al., 2009; Singh et al., 2017). Figure 1.1 shows media optimization and sterilization task

1.2.1. Media Composition Optimization Approaches

1.2.1.1. Process System Engineering (PSE)

Process system engineering (PSE) is current used good technique and tool and it needs to understand its basic limitations. This approach would be good to implement in fermentation processes with special focus on mathematical modeling approaches. The paper ends with a number of future perspectives in this area: several potential solutions are proposed to facilitate the future use of PSE methods and tools in fermentation process development, operation and optimization (Rizwan et al., 2019). This approach was applied in designing of ionic liquid media and also designing of integrated biorefineries as two main research areas. For this approach implementation in fermentation process, first needs for development and advances of novel computational tool with capabilities for optimization of media. This can apply in practical results verifications (Fasahati et al., 2019). Normally in introduction of optimization approach application to ionic liquid media is discussed with some major challenges. But this approach can make possible to provide more reliable for optimal system design with minimizing of production cost of liquid media and also in environmental impacts with this media utilization (Chemmangattuvalappi et al., 2020). Researchers have put their efforts for recent development in PSE development in integrated biorefineries area development and numbers of different value added products are reported to biosynthesize from biorefineries approach with aided of application of PSE tool. This showed the aims of enhanced sustainable manner performance at environmental and social level/ influences (Rizwan et al., 2019; Chemmangattuvalappi et al., 2020).

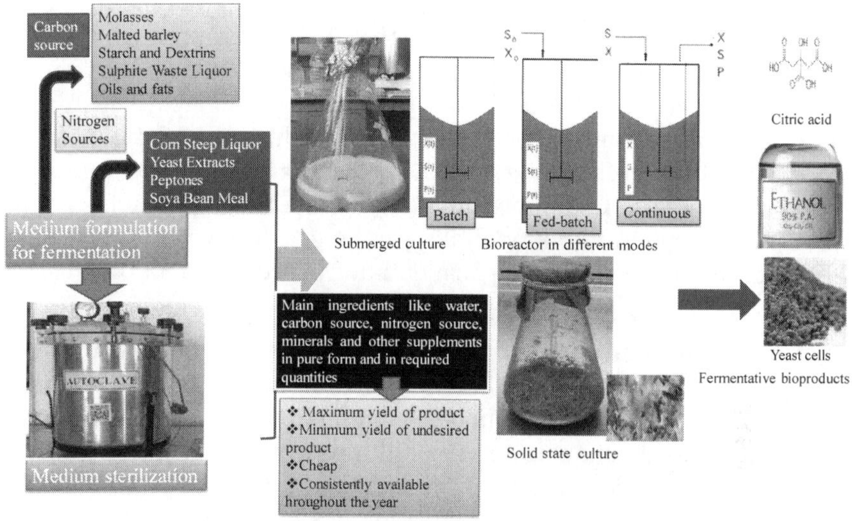

Figure 1.1. - Fermentative media designing from different carbon and nitrogen substrates shown for metabolites production via using different modes of fermentation processes.

In current period, biofuel synthesis is discussed from utilization lignocellulosic biomass or its waste residues with applying of efficient fermentation platform/ bioprocess and it is sustainable mode of energy development that can provide alternative to fossil fuel. In this regard PSE approach was found effective tool with utilization of computer-based tools/ methods. This approach helps in designing the process simulation and optimization of fermentative parameters (Darkwah et al., 2018). Now, utility of PSE is found to design the economic modes of biorefineries process. This task can be achieved via development of simulation approaches with integration with existing mature PSE tools. Finally optimal nature of traditional refineries (i.e., Aspen plus Current plus) can achieve. Some current unit operation models showed their lacking ability for describing the unsteady state fermentation processes or linkage of unsteady state fermentation with in situ separation and also optimization of fermentative process such as yield/ productivity as competing factors (Geraili et al., 2014). Now, using of novel architecture of commercial tools, Aspen Plus and MatLab tools effort can help in development of techniques for best way simulation time-dependent fermentation without or with in situ separations for process designs analysis and optimization of operating conditions in fermentation. .In recent years, traditional batch fermentation process are simulated with in situ separation decouple approach and these are interdependent steps in a separate steady state bioreactor/ reactor with achieving of equilibrium separation of the final fermentation broth that contained metabolites (Jiménez-González et al., 2010).

A typical mechanistic system was applied for ordinary different equations (ODEs) that describe a batch fermentation process, not fit the standard built-in power law reactions kinetic model in Aspen Plus. So this challenge can solve by using a novel platform that exhibits the linkage the batch reactors to a FORTRAN user kinetics subroutines via incorporating the ODEs. This approach can combine with components substitution with capacity to simulate non-databank components (Vázquez-Ojeda et al., 2013). And these can utilize to simulate an unsteady state batch and in situ gas stripping process. Final form of this model system can predict the product synthesis profile to the sensitive to the gas flow rate that did not report in previous steady state. This effort can demonstrate the importance of linking a time dependent fermentation model that can help to fermentation environment and also designing and analyzing the fermentation processes (Kim and Reed 2014).

In recent years, researchers developed the novel platforms that are linking to genetic algorithm multi-objective and single-objective optimizations in the software Matlab to unsteady state batch fermentation simulation in Aspen Plus. This effort was found via a component object module communication platform that can help in optimization of operational conditions in a typical batch fermentation processes (Vázquez-Ojeda et al., 2013; Kim and Reed 2014). There are two major contributions was found before concentration of sugars from liquid fraction of a typical lignocellulosic hydrolysis steps/ processes. This can be needed with high quantity of initial sugar concentration. And typical fermentation process can integrate with an in situ separation process with optimized fermentation processes with best performance (Jiménez-González et al., 2010; Darkwah et al., 2018). And this frame-work can be analysized by fermentation experimentalist that can use the full site of PSE tools and method for integration to biorefineries and refineries via working a decision support tool to guide the designing, analysis and optimization of fermentation-based biorefineries (Kim and Reed 2014; Darkwah et al., 2018).

1.2.1.2. Genetic Algorithm for Media

There are many approaches of fermentative media optimization and metaheuristics genetic algorithm is found as best approach and this GA is based on the natural selection process and this approach is fallen under the umbrella category of evolutionary algorithm (EA). This GA approach utility is found in generation of high-quality solution for search and optimization problem. And this GA is dependent on bio-oriented operators such as selection, crossover and mutation. This GA can still suffer from some downsides and it must to improve for attaining good/ greater control of exploitation and exploration concern (Albadr et al., 2020).

This can create a new population and randomness involvement and it can also happen in the population at the solution initialization. As we are aware that mutation process is imposed upon the new chromosomes and it can prevent the achievement of an optimal solution/ results. In recent years, this tool is presenting a new GA technique for media optimization for fermentation and it can center on the natural selection theory and this can aims to improve the control of exploitation and exploration (Contreras-Bolton et al., 2015). The new proposed genetic algorithm is based on natural selection theory (under name of GABONST). And in this context two assessment of the GABONST are explored on the basis of (1) application of fifteen renowned benchmark test functions and also the comparison of the results with the conventional/ traditional GA. This is also known as enhanced ameliorated teaching-based optimization (EATLBO) (Zhang et al., 2019). And (2) application of GABONST in languages identification (LID) showed via integrating the GABONST with extreme learning machine (ELM) under the name of GOBONST-ELM. Now, ELM is one of most useful learning models for carrying out classification and regression analysis. Then generated results are shown to carryout ground upon the LID dataset. And this GA can be derived from eight separate languages (Wong et al., 2020).

The GABONST algorithm approach is shown to capable of production of good quality of solution that can show better control of the exploitation and exploration as compared to traditional/ conventional genetic algorithm (GA) EATLBO, Bat and Bee algorithm under the assessment of statistical tools/methods. In addition benefits from generated results have reported the GABONST-ELM-LID tool that showed effective performance and accuracy up to 99.4% (Zhang et al., 2019; Wong et al., 2020).

In media optimization context, sponges are used optimization of amino acids concentration/ compositions in a commercially available basal cell culture medium that can increase/ enhance the metabolic activity of cells of marine sponge *Dysidea etheria*. And this sponge is rich source of novel natural products/ bioproduct. And production of cell culture can provide best option for supply these natural compound and these are not currently found in sponge cell lines (Camacho-Rodríguez et al., 2015). But there is lack of understanding about the precise conditions and nutritional requirements that can require sustaining sponge cells in vitro conditions. And still it needs to be defined sponge-specific nutrient medium formulations (López-Rosales et al., 2015). In this published work, GA approach utility is found optimization of amino acid composition that is required in commercially available basal cell culture medium. In this regard, four generations of GA were performed in vitro in wet laboratory conditions (Munroe et al., 2019).

And optimized medium composition combination was selected for further evaluation process. And it was whenever compared to basal medium control then this approach of optimization can find two fold enhance in metabolic activity. And GA tool can choose the parameters without the need for details knowledge on all possible interactions (Camacho-Rodríguez et al., 2015; Munroe et al., 2019). Another research paper explored the characteristics of high mobility of Flying Ad Hoc Network (FANETS) nodes and it was combined with the features of Topology-based class routes protocol on-demand search. This approach is a genetic algorithm-based routing (GAR) protocol and is proposed for FANETs, based on improved GA for FANETs route search (Wei et al., 2021).

This technique takes into account the link stability, link bandwidth node energy and others factors. Next GAR can improve the selection, crossover and variation operators of the GA (Abualigah et al., 2020). And GAR finally found to be planned for optimization state from the communication initiating node to the destination nodes in quick manner via using a smaller cost. And experimental results from GAR approach is highly improved the throughput via reduced the delay and improved stability of the network and is more suitable for FANETs tasks (Abualigah et al., 2020; Wei et al., 2021).

1.2.1.3. Artificial Neural Network for Media

In current year, artificial neural network (ANN) tool is discussed for fermentation process for their optimized media components and concentration that can help to improve the yield and titer of any particular bio-products and it is very commonly used technique in media optimization for biosynthesis of different products in bioprocesses and also enzyme synthesis. In this context, some studies showed the descriptive and comparative approaches of response surface methodology (RSM) and artificial neural network (ANN) (Chen et al., 2015).

Further numerous reports description was shown for better quantity of enzyme and biomolecules biosynthesis via application of ANN technique via compared to RSM technique. Except microbiology, ANN approach can use in different scientific optimization process in successful manner. And ANN introduction in review tasks is focused on some studies that is used for microbial fermentation processes for biosynthesis of biomolecules in descriptive forms via compared to RSM for same media optimization task (Dhanarajan et al., 2014). And many research groups is found to use the ANN technique and is mathematical and statistical method or modeling approach with exhibiting of wide range of application in biotechnological processes. These processes are optimized processes and also optimized media in bioprocess that used for various biomoelcules biosynthesis in biological system. This approach further showed its efficiency in increasing enzyme yield in different microbial sources (Peng et al., 2014).

And ANN approach is found to consist of dense interconnected computing units in simple models for complex neurons in biological system. And ANN tool is biological manner motivated and also showed mimics human brain. Further ANN tool is consisted of a large network of neuron, connected with connection link. This ANN tool can exhibit to contain an activation function for elucidating the output. Normally nonlinear activation functions are reported and also are sigmoid and step function for uses (Yadav et al., 2013). ANN tool are trained by experience and it can be applied for a new input to the network and can generalize from past experiences that can produce a new result. The simple structure of ANN tool is normally is made up of an input layer and an output layer. And application of algorithm can mimic the processes of real neurons and the network can find to learn to solve many types of problems. Next, this ANN toll can provide a mathematical alternative to the quadratic polynomial for representing data that derived from statistical designed experiments (Yadav et al., 2013; Peng et al., 2014).

And we know the capability of ANN tool that can handle large amount of data to approximate functions to any desired degree of accuracy and this tool can make it attractive as empirical model. And ANN tool is used in different filed of science that includes physics, chemistry and management. In these fields also this ANN utility is found in predicting the secondary structure of global proteins (Yadav et al., 2013; Chen et al., 2015). In maximum efforts for fermentative media optimizations and their benefits, microbial fermentation process are found for biosynthesis of bioemulsifiers agent and exopolysaccharides and are found more industrially important biomolecules and these have shown their application in food, cosmetics and pharmaceutical industries/areas. And these both compound are biosynthesized in extracellular and secondary metabolites via application/ utilization of appropriate microbial system (Dhagat et al., 2021).

In this published work in recent years, two statistical optimization tools (such as simulated annealing and ANN tools) are reported and are used for optimized modes of these bioemulsifier and exopolysaccharides biosynthesis in thermophilic *Brevibacillys borstelensis* microbial system (Tong et al., 2021). These bioprocess variables were glucose concentration, monosodium glutamate, yeast extract and magnesium sulfate in production media and concentration of bioemusifiers, exopolysaccharides and biomass, studied as responses of this process. For this bioprocess, the best function values for the three responses are obtained via simulated annealing approaches (7.14 $g.L^{-1}$, 17.9 $g.L^{-1}$ and 3.3 $g.L^{-1}$) and also three responses generated by ANN tools predictions (6.1 $g.L^{-1}$, 17.5 $g.L^{-1}$ and 2.5 $g.L^{-1}$) (Vo et al., 2020). This research study showed its efficiency of statistical optimization tools via comparing to the experimental approaches for media optimization of bioprocesses with maximum yield and titer of any metabolites. And these applied techniques can reduce the time and cost of any bioprocesses that used for metabolites synthesis (Ma et al., 2019).

And in recent years, there is huge and cheap sources of agricultural residues and these are associated the many problems for their utilization and reduction in quantity that can create environmental issues. And agricultural waste is found to contain high amount of carbon and nitrogen sources/ substrates and these are utilized for suitable substrates for biosynthesis of bioemulsifier and exopolysaccharides in fermentation processes (Dhagat et al., 2021; Tong et al., 2021). The concentration of medium components can be obtained from optimal media compositions and this can be useful for agricultural wastes as efficient substrates. This can lead to biosynthesis of these two industrially important compounds via using sustainable substrates sources with proceeding of environmental remediation (Ma et al., 2019; Dhagat et al., 2021).

In this modern era, there is looming of energy crisis due to increasing rate of depletion of fossil fuels resources and environmental concern from greenhouses gas emissions. These concerns can solve by developing of sustainability energy sources with its increasing needs in future period. These have secured global attention. And more numbers of research works is currently focused towards the development of renewable sources of energy and this energy can utilize more waste matters and also their availability and environmental friendliness (Beiroti et al., 2019). In this context, numbers of biofuel biosynthesis via using of many bioprocesses that is controlled by several process parameters including pH, temperature and substrates concentration. And in recent periods, many approaches used in improvement biofuel biosynthesis process and these used many robust process model that can accurate manners with relation to effect of input variable to the process output (Liyanaarachchi et al., 2021). ANN tool is emerged as a best tool for modeling complex, non-linear processes. ANNs are found to apply in the prediction of various processes that can be useful for virtual experimentations. These can potentially enhance bioprocess research and development (Yahya et al., 2015).

This published work study explored the recent findings on the application of ANN that was used for modeling and optimization of bioH2, biogas, biodiesel bioethanol and also microbial fuel cells via uses of microbial technology. In additional to comparative studies on modeling efficiency of ANN tool and other techniques (i.e., RSM application with more description) are found. Finally it can conclude that ANNs utility in modeling and optimization tasks in biofuel process development (Sewsynker-Sukai et al., 2017; Liyanaarachchi et al., 2021).

1.3. Metabolic Engineering and Others Approaches for Metabolites

As we know that L-tryptophan amino acid is aromatic nature of amino acid with widely used in medicine, food and animal feed. And recent year, microbial processes for L-tryptophan biosynthesis are explored with application of metabolic engineering tools that can be found as a sustainable, cost-effective and eco-friendly process routes (Liu et al., 2019). This can be compared to various routes of chemical synthesis for this amino acid. In particular, metabolic pathway engineering can help in enhancement of products titers via inactivating / blocking the competing pathways and this can increase the intracellular level of essential precursor and over-expression of rate-limiting enzyme steps (Liu et al., 2017). This approach is based on the route of the L-tryptophan biosynthesis pathways and many reviews and research works are shown to present a systematic and details summery of the contemporary metabolic engineering that employed for L-tryptophan biosynthesis. There are many other engineering approaches except metabolic engineering approach that can improve the biosynthesis of L-tryptophan pathways. Metabolic engineering tool helps in modification of sources of carbon uptake, byproducts formation, and key regulatory factors during polyhydroxybutyrate and L-tryptophan biosynthesis pathway (Liu et al., 2016; Liu et al., 2019).

These factors are explored in details with fermentation bioprocess optimization strategies that helped in L-tryptophan overproduction in sustainable manner with its more descriptive form. This published paper explored with its conclusion that high L-Tryptophan production microbial can generate from this tool that can overproduction of these metabolites in fermentation processes (Liu et al., 2017; Liu et al., 2019).

1.3.1 Metabolites from Microbial Fermentation

Most of metabolites are found to biosynthesize via using solid sate fermentation (SSF) under microbial fermentation and this approach helps in taking place in absence/ near absence of free water and this situation can be very close to natural environment for some selected microbial systems/ strains especially for fungal species with its natural adaptation for this condition (Manan et al., 2017). This SSF is utilized from ancient time at worldwide and also known to apply in the East for traditional manufacture of fermented foods and also in the west for mould-ripened cheese. This (SSF) approach is explored as system with growth of selected microbial strains using a solid material with low moisture content (Ashok et al., 2017). SSF is also found to potentially important methodology and technique in biotechnology. In recent period, this SSF approach is utilized as economically viable and practically acceptable technology for large-scale bioconversion and biodegradation processes. And development of sustainable SSF and others bioprocess technology can be an option with emerging and multidisciplinary field favorable approach (Jiménez-Peñalver et al., 2017).

This approach can find possible application to production of enzymes, chemicals, bioethanol and pharmaceutical. SSFs showed multiples advantages over conventional submerged fermentation (SMF) and it is found to use simple and inexpensive substrates. This can eliminate the need of solubilization of nutrient from within solid substrates and also eliminates the need of rigorous control for this process that required for many others processes including many parameters during others fermentation (Jiménez-Peñalver et al., 2017; Ashok et al., 2017). Product yield is reported in higher quantity with lower energy requirement and also less amount of waste water production and comparatively easier mode of end product recovery. SSF can provide in flexibility in term of the raw material utilization and also their capability for production of various nature of value-added products (Manan et al., 2017; Ashok et al., 2017).

In many reports, industrial fermentation is found via intentional uses of fermentation and appropriate microbial/ others biological system such as bacteria, fungi and also eukaryotic cells like Chinese Hamster Ovary (CHO) and insect cells. These biological systems have contributed to make the many products useful to humans. As we seen that many fermented products have shown many application including food and general industries (Parapouli et al., 2020). And some commodity chemicals such as acetic, citric acid and ethanol are made by fermentation process. And the rate of fermentation depends on concentration/ effective nature of microbial system, cells and cellular components, enzymes, temperature and pH and these are important factors in aerobic fermentation with oxygen need. And product recovery is frequently used / involved the concentration of dilute solution (Papalexandrato et al., 2011).

And nearly all commercial products (such as enzymes (lipase, invertase and rennet) are produced via fermentation with utilization of genetically modified microbial system. Single cell protein and baker yeast and also lactic acid bacteria is used as starter culture for cheese making process where biomass production was itself the main objective of fermentation (Papalexandrato *et al.*, 2011; Parapouli *et al.*, 2020).

There are four types utility of fermentation processes are found and these are production of biomass (viable cell materials), extracellular metabolites (chemical compounds), intracellular components (enzyme and other proteins) and also transformation of substrates (including products itself). These processes are not need to disjoint from each other but these can provide a framework for understanding the difference in approach. In these processes, organisms can be bacteria, yeast, molds, algae, animal cells and plant cells with special consideration in fermentation such as dissolve oxygen, nutrient level and temperature and some metabolites can find in table 1.1 (Kumar et al., 2017).

Table 1.1 - Metabolites with high yield and titer in different fermentative processes

Fermentative media	*Products/ Services*	*Reference*
Original and modified/ complex media conditions of fermentation	Maximum quantity of this fungal biomass 22and 33 g.L-1 respectively in Original and modified/ complex media	Klaic et al.,2015
Suitable concentration of compositions in fermentation media	Promoted the growth of Phoma species (such as P. loveata) as potential fungal species (pathogen of potatoes plants)	Rohlfs et al., 2011
Recombinant strain Escherichia coli and Saccharomyces cerevisiae in fermentation processes	It helped in production recombinant protein via using these strain like insulin	Rezende et al., 2012
Commercially available basal cell culture medium is optimized by GA tool can choose the parameters	It can increase/ enhance the metabolic activity of cells of marine sponge Dysidea etheria	Munroe et al., 2019
Media optimizations in microbial fermentation process	High yield and titer of biosynthesis of bioemulsifiers agent and exopolysaccharides	Dhagat et al., 2021
Metabolic engineering tool helps in modification of sources of carbon uptake, byproducts formation, and key regulatory factors	This approach improved the yield and titer for polyhydroxybutyrate and L-tryptophan biosynthesis pathway in sustainable manner	Liu et al., 2019
Solid sate fermentation (SSF) under suitable microbial system	This approach is best and possible application to production of enzymes, chemicals, bioethanol and pharmaceutical	Jiménez-Peñalver et al., 2017; Ashok et al., 2017

1.4. Media Sterilization

For fermentation process to complete in successful manner and efficient way, fermentative media sterilization is necessary with need to sterilize the nutrients based on renewable (maize and soya waste residues and also complex organic matters) and non-renewable raw material (sugarcane juice, food grains, and others plant products). As we know that most industrial enzymes are secreted by many types of microbial into the fermentation broth and these enzymes can help to breakdown the complex sources of carbon and nitrogen sources. Batch, fed-batch and continuous modes of fermentation process are common process (Senz et al., 2019).

So, prevention of any microbial contamination primarily needs for much industrial food or other nutrients fermentation process. Media sterilization can destruct/ remove all the form of microbial life from the aqueous feedstock. In industrial fermentation, many components such as vessels, pipe works, media, inlet air and exhaust air are necessary to sterilize via combination of wet-heat and filtration approach. And wet-heat approaches are found less expensive and more effective approach than heat-heat approach and this can employ commonly in industrial fermentation and also completing works in laboratory and health care (Corzo-Martínez et al., 2012).

These processes showed their efficiency for destroying unwanted microbial system that harms the fermentation processes with their end products. The wet-heat sterilization approach is more efficient with capability to kill all type of microbial system including bacterial vegetative cells, and spores. And these contaminants in condition of wet-heat sterilization in autoclave vessel are found with an atmosphere of saturated steam. In all biotechnological process media sterilization is done with applying heat and to a lesser extent of sterilization can be achieved by other means such as physical method, chemical treatment and radiation (Senz et al., 2015) and can be seen in table 1.2.

Heat sterilization is widely used sterilization tool/ technique and this can define as any process that achieves effective killing or elimination of transmissible agent such as fungi, bacteria, viruses and prions from a surface, food, medication or biological culture medium. And industrial media are done for complete and effective sterilization with steam with destroying the indigenous microbial system before inoculation of fermentative media with specific microbial cells in huge number/ population (Russo et al., 2017). There is also need of biological validation of each sterilization cycle and it can minimize the impractical of an overkill approach and these are commonly employed on the basis that can alter the heat-sensitive to less detrimental nutrients without survival of indigenous microbial system. It has shown the effectiveness of heat destruction of microbial system and it exhibits probability function amenable to calculation (Salvetti et al., 2017).

A computer device has been applied for programming to calculate the on-line heat input as F_0 value during sterilization of fermentative media sterilization in stirred bioreactors. The accumulation of F_0 value is calculated verbally o bioreactor operators via communication controller with voice synthesizers (Boeck et al., 1988). Next, batch sterilization is found most commonly approach for liquid media sterilization. This mode of sterilization can heat to sterilization temperature via introducing steam into the coils or double jacket of the vessel and also alternative modes with steam injection directly into the medium or vessel that need to sterilize/ heated electric power energy modes (Sharma *et al.*, 2018).

Direct steam injection utility can allow with dilution of the medium via condensate. It is typically added to 10 to 20% to the liquid volume and quality of steam. These can be sufficient with high heat capacity to avoid contamination of the bioprocess substrates via metal ions or organics. Further, depending on the process temperature and contact time, heat applicant can frequently lead to change in certain starting material with shown of adverse affect to course of fermentation (Memarzadeh et al., 2014). Some of these are caramelization of sugar solution, denaturing of protein (as used as nitrogen sources), inactivation of numerous vitamins or another substance (essential to microbial growth), reaction of aldo sugars with amino acid or other substance (also require for microbial growth) (Taylor *et al.*, 2016).

These adverse effects due to various reactions with nutrients, can be avoided by sterilization the individual components of nutrients medium separately or occasionally pH value regulation. Certain nutrient such as the carbon sources and also the precursor solution are fed in continuous modes during bioprocess and it needs to sterilize before reach to bioreactor. This process is accomplished with the help of instantaneous heaters and it needs to designed with such dimensions and desired lethal effect that need to achieve at a tolerable temperature in a acceptable time (Zhang et al., 2011; Noorman et al., 2018). And high temperature short time (HTST) mode of heat sterilization can minimize the much time loss at specific values of temperature. And 130 to 140^{0}C are used in HTST process in most process with applying of steam injection (Groen et al., 2015; Noorman et al., 2018).

Table 1.2- Sterilizing agents used in bioprocess and health-care

Sterilization agents	*Microbial agents affected*	*References*
1-2.0%–2.3% Glutaraldehyde	Effective in eliminating H. pylori	Cronmiller et al., 1999; Kaneko et al., 1993)
Glutaraldehyde (with and without active peracetic acid)	Effective in eliminating H. pylori	Cronmiller et al., 1999, Roosendaal et al., 1993
Chlorine at ~1 ppm in alfalfa seeds or sprouts, or beef carcass surfaces	Eliminating approximately 4 log10 of E. coli O157:H7 within 1 minute	Taormina et al., 1999; Castillo et al., 1999
Sodium hypochlorite (800 ppm free chlorine),/ a phenol-based product (14.7% phenol diluted 1:256 in tapwater)	Effective in interrupting transfer of a human rotavirus from stainless steel disk	Sattar et al., 1994; Kaneko et al., 1993

Sterilization agents	*Microbial agents affected*	*References*
Chlorine dioxide gas as surface disinfection	It is found for B. anthracis spores inactivation	Weber et al., 2004
From phenolics to quaternary ammonium compounds	Used for or low-level disinfection	Jordan et al.,1997
Chemicals such as sodium bisulfate or glycine is used for instruments and ventilation	Used for the medical community with good important and effective influences	Weber et al., 2004; Jordan et al.,1995
Effectiveness of spraying the fabric with 3% hydrogen peroxide is shown	Used on surfaces in a health-care facility as surfaces cleaner, effective for health-care-associated infections.	Griffith et al., 2000
Hydrogen Peroxide accounts several properties of germicidal effectiveness as bactericidal, virucidal, sporicidal, and fungicidal properties	Attest due to producing destructive hydroxyl free radicals attack membrane lipids, DNA, and other essential cell components.	Turner et al., 1983; Block et al., 2001; Sattar et al., 1998.

1.5. Conclusion

This chapter explored the importance of media compositions that are important for different types of fermentation processes. Nowadays, fermentation is eco-friendly processes and is applied for primary and secondary metabolites. As we discuss the importance of media compositions and also media sterilization that can make the fermentation process success with more efficient for metabolites biosynthesis in term of higher titer and yield with lesser quantities of byproducts. This chapter focused the different approaches such as process system engineering, genetic algorithm and also artificial neural network for media composition optimization tasks. We also explored the overall performance of fermentation processes with higher yield for metabolite via application of metabolic engineering. Media sterilization task with efficient approach (wet steam method) can help to completing of fermentation processes with sterilized conditions. This can help to commercial people for designing of media formulation for their biotechnological bioproducts

1.6. Abbreviations

ANN: Artificial neural network; **CHO:** Chinese Hamster Ovary; **EA:** Evolutionary algorithm; **EATLBO:** Enhanced ameliorated teaching-based optimization; **ELM:** Extreme learning machine; **FANETS:** Flying Ad Hoc Network; **FORTRAN:** Formula Translation computer programming language; **GA:** Genetic algorithm; **GABONST:** Genetic algorithm is based on natural selection theory; **GAR:** Genetic algorithm-based routing; **LID:** Languages identification; **ODEs:** Ordinary different equations; **PSE:** Process system engineering; **RSM:** Response surface methodology; **SMF:** Submerged fermentation; **SSF:** Solid sate fermentation;

References

Abualigah, L., & Diabat, A. (2020). A comprehensive survey of the Grasshopper optimization algorithm: results, variants, and applications. Neural Comput. Appl, 32, 15533–15556. doi: 10.1007/s00521-020-04789-8.

Albadr, M.A., Tiun, S., Ayob M., & AL-Dhief, F. (2020). Genetic Algorithm Based on Natural Selection Theory for Optimization Problems. Symm, 12(11), 1758. https://doi.org/10.3390/sym 12111758

Ashok, A., Doriya, K., Rao, D.R.M., & Kumar, D.S. (2017). Design of solid-state bioreactor for industrial applications: An overview to conventional bioreactors. Biocatalyst Agric Biotechnol, 9, 11–18.

Beiroti, A., Aghasadeghi, M.R., Hosseini, S.N., & Norouzian, D. (2019). Application of recurrent neural network for online prediction of cell density of recombinant Pichia pastoris producing HBsAg. Prepare Biochem Biotechnol, 49, 4, 352-359.

Berendsen, B.J.A., Gerritsen, H.W., Wegh, R.S., Lameris, S., van-Sebille, R., Stolker, A.A.M., & Nielen, M.W.F. (2013). Comprehensive analysis of ßlactam antibiotics including penicillins, cephalosporins, and carbapenems in poultry muscle using liquid chromatography coupled to tandem mass spectrometry. Anal. Bioanal. Chem, 405 (24), 7859–7874.

Block, S.S. (2001). Peroxygen compounds. In: Block SS, ed. Disinfection, sterilization, and preservation. Philadelphia: Lippincott Williams & Wilkins, 185-204.

Boeck, L.V.D., Wetzel, R.W., Burt, S.C., Huber, F.M., Fowler, G.L., & Alford, J.S. (1988). Sterilization of bioreactor media on the basis of computer calculated thermal input designated as Fo. Journal of Industrial Microbiol, 3, 305-310.

Camacho-Rodríguez, J., Cerón-García, M.C., Fernández-Sevilla, J.M., & Molina-Grima, E. (2015). Genetic algorithm for the medium optimization of the microalga Nannochloropsis gaditana cultured to aquaculture. Bioresour Technol, 177:102–109.

Castillo, A., Lucia, L.M., Kemp, G.K., & Acuff, G.R. (1999). Reduction of Escherichia coli O157:H7 and Salmonella typhimurium on beef carcass surfaces using acidified sodium chlorite. J. Food Prot, 62, 580-4.

Chemmangattuvalappil, N.G., Ng, D.K.S., Ng, L.Y., Ooi, J., Chong, J.W., & Eden, M.R.A. (2020). Review of Process Systems Engineering (PSE) Tools for the Design of Ionic Liquids and Integrated Biorefineries. Processes, 8, 1678. https://doi.org/10.3390/pr8121678

Chen, F., Li, H., Xu, Z., Hou, S., & Yang, D. (2015). User-friendly optimization approach of fed-batch fermentation conditions for the production of iturin A using artificial neural networks and support vector machine. Electr J Biotechnol, 18, 273–280. https://doi.org/10.1016/j.ejbt. 2015.05.001

Contreras-Bolton, C., & Parada, V. (2015). Automatic Combination of Operators in a Genetic Algorithm to Solve the Traveling Salesman Problem. PLoS ONE, 10, e0137724.

Corzo-Martínez, M., Ávila, M., Moreno, F.J., Requena, T., & Villamiel, M. (2012). Effect of milk protein glycation and gastrointestinal digestion on the growth of bifidobacteria and lactic acid bacteria. Int. J. Food Microbiol, 153, 420–427.

Cronmiller, J.R., Nelson, D.K., Jackson, D.K., & Kim, C.H. (1999). Efficacy of conventional endoscopic disinfection and sterilization methods against Helicobacter pylori contamination. Helicobacter, 4, 198-203.

Darkwah, K. (2018). Application of process systems engineering tools and methods to fermentation-based biorefineries. Theses and Dissertations--Chemical and Materials Engineering, 83. https://uknowledge.uky.edu/cme_etds/83

Dhagat S., & Jujjavarapu S.E. (2021). Simulated annealing and artificial neural network as optimization tools to enhance yields of bioemulsifier and exopolysaccharides by thermophilic Brevibacillus borstelensis. J Environ Chem Eng, 9(4), 105499. https://doi.org/10.1016/j .jece.2021.105499

Dhanarajan, G., Mandal, M., & Sen, R. (2014). A combined artificial neural network modeling particle swarm optimization strategy for improved production of marine bacterial lipopeptide from food waste. Biochem Eng J, 84, 59–65. doi:10.1016/j.bej.2014.01.002

El-Elimat, T., Zhang, X., Jarjoura, D., Moy, F.J., Orjala, J., Kinghorn, A.D., Pearce, C.J., & Oberlies, N.H. (2012). Chemical diversity of metabolites from fungi, cyanobacteria, and plants relative to FDA-approved anticancer agents. ACS Med Chem Lett, 3, 645–649. doi: 10.1021/ml300105s.

Fasahati, P., Wu, & W., Maravelias, C.T. (2019). Process synthesis and economic analysis of cyanobacteria biorefineries A superstructure-based approach. Appl. Ener, 253, 11362.

Fatima, S., Rasool, A., Sajjad N., Bhat, E.A., Hanafiah, M.M., & Mahboob, M. (2019). Analysis and evaluation of penicillin production by using soil fungi. Biocataly Agricult Biotechnol, 21, 101330.

Geraili, A., Sharma, P., & Romagnoli, J.A. (2014). A modeling framework for design of nonlinear renewable energy systems through integrated simulation modeling and metaheuristic optimization: Applications to biorefineries. Comput. Chem. Eng, 61 (0), 102-117.

Griffith, C.J., Cooper, R.A., Gilmore, J., Davies, C., & Lewis, M. (2000). An evaluation of hospital cleaning regimes and standards. J. Hosp. Infect, 45, 19-28.

Groen D.J., Noorman H.J., & Stankiewicz A.I. (2015). Improved method for aerobic fermentation intensification, Proceedings of Sustainable (Bio)Chemical Process Technology Incorporating 6th Int. Conf. Process Intensification , P.J. Jansens, A.I. Stankiewicz and A. Green, Cranfield Group Limited, pp 105–112.

Guzmán-Chávez, F., Zwahlen, R.D., Bovenberg, R.A.L., & Driessen. A.J.M. 2018. Engineering of the Filamentous Fungus *Penicillium chrysogenum* as Cell Factory for Natural Products. Front. Microbiol, 9, 2768. doi: 10.3389/fmicb.2018.02768

Jiménez-González, C., & Woodley, J. M. (2010). Bioprocesses: Modeling needs for process evaluation and sustainability assessment. Comput. Chem. Eng, 34 (7), 1009-1017.

Jiménez-Peñalver, P., Gea, T., Sánchez, A., & Font, X. (2017). Production of sophorolipids from winterization oil cake by solid-state fermentation: Optimization, monitoring and effect of mixing. Biochemical Engineering Journal, 115, 93–100. https://doi.org/10.1016/j .bej.2016. 08.006

Jordan, S.L. (1995). The correct use of glutaraldehyde in the healthcare environment. Gastroenterol. Nurs, 18, 143-5.

Jordan, S.L.P., Russo, M.R., Blessing, R.L., & Grab, L.A. (1997). Glutaraldehyde safety: inactivation and disposal. Am. J. Infect. Control, 25, 154-55.

Kaneko, H., Mitsuma, T., Kotera, H., Uchida, K., Furusawa, A., & Morise, K. (1993). Are routine cleaning methods sufficient to remove Helicobacter pylori from endoscopic equipment Endosc, 25, 435.

Kim, J., & Reed, J.L. (2014). Refining metabolic models and accounting for regulatory effects. Curr. Opin. Biotechnol, 29, 34-38.

Klaic, R., Sallet, D., Kuhn, R.C., Salbego P., Centenaro G.S., Jacques, R.J.S., Guedes, V.C., Jahn S.L., & Mazutti M.A. (2015). Optimization of fermentation media for the growth of a fungus used as bioherbicide. Conference: XX Congresso Brasileiro de Engenharia Química.DOI:10.5151/CHEMENG-COBEQ2014-1447-19247 146606.https://www.researchgate.net/publication/300663909_Optimization_of_Fermentation_Media_For_The_Growth_Of_A_Fungus_Used_As_Bioherbicide

Kumar, V., Chhabra, D., & Shukla, P. (2017). Xylanase production from *Thermomyces lanuginosus* VAPS-24 using low cost agro-industrial residues via hybrid optimization tools and its potential use for saccharification. Bioresour Technol, 243, 1009-1019. doi: 10.1016/j.biortech. 2017.07.094.

Liu, L., Bilal, M., Luo, H., Zhao, Y., & Iqbal, H.M.N. (2019). Metabolic Engineering and Fermentation Process Strategies for L-Tryptophan Production by Escherichia coli. Process, 7, 213. https://doi.org/10.3390/pr7040213

Liu, L., Chen, S., & Wu, J. (2017). Phosphoenolpyruvate: Glucose phosphotransferase system modification increases the conversion rate during L-tryptophan production in *Escherichia coli*. J. Ind. Microbiol. Biotechnol, 44, 1–11.

Liu, L., Duan, X., & Wu, J. (2016). Modulating the direction of carbon flow in Escherichia coli to improve L-tryptophan production by inactivating the global regulator FruR. J. Biotechnol, 231, 141–148.

Liyanaarachchi, V.C., Nishshanka, G.K.S.H., Sakarika, M., Nimarshana, P.H.V., Ariyadasa, T.U., & Kornaros, M. (2021). Artificial neural network (ANN) approach to optimize cultivation conditions of microalga Chlorella vulgaris in view of biodiesel production. Biochem Eng J, 73, 108072.

López-Rosales, L., García-Camacho, F., Sánchez-Mirón, A., Contreras-Gómez, A., & Molina-Grima, E. (2015). An optimisation approach for culturing the shear-sensitive dinoflagellate microalgae in bench-scale bubble column photobioreactors. Bioresour Technol, 197, 375–382.

Ma, S., Tong, L., Ye, F., Xiao, J.S., Benard, P., & Chahine, R. (2019). Hydrogen purification layered bed optimization based on artificial neural network prediction of breakthrough curves Int J Hydrogen Energy, 44, 5324-5333.

Manan, M.A., & Webb, C. (2017). Design aspects of solid-state fermentation as applied to microbial bioprocessing. J Appl Biotechnol Bioeng, 4(1), 511-532. DOI: 10.15406/jabb.2017 .04.00094

Memarzadeh, F. (2014). Adding amines to steam for humidification. Chem. Health Saf, 21, 4, 5–17. https://doi.org/10.1016/j.jchas.2014.01.002

Miao, L., Kwong, T.F.N., & Qian, P.Y. (2006). Effect of culture conditions on mycelial growth, antibacterial activity, and metabolite profiles of the marine-derived fungus Arthrinium c.f. saccharicola. Appl Microbiol Biotechnol, 3,1063–1073. doi: 10.1007/s00253-006-0376-8

Molen, K.M.V., Raja, H.A., El-Elimat, T., & Oberlies, N.H. (2013). Evaluation of culture media for the production of secondary metabolites in a natural products screening program. AMB Expr, 3, 71. doi: 10.1186/2191-0855-3-71

Munroe, S., Sandoval, K., Martens, D.E., Sipkema, D., & Pomponi, S.A. (2019). Genetic algorithm as an optimization tool for the development of sponge cell culture media. In Vitro Cell.Dev.Biol.Animal, 55, 149–158. https://doi.org/10.1007/s11626-018-00317-0

Noorman, H.J., van-Winden, W., Heijnen, J.J., & van der Lans R.G.J.M. (2018). Intensified Fermentation Processes and Equipment, in Intensification of Biobased Processes, pp 1-41. DOI: 10.1039/9781788010320-00001, eISBN: 978-1-78801-032-0.

Osama, K., Somvanshi, P., Pandey, A.K., & Mishra, B.N. (2013). Modelling of nutrient mist reactor for hairy root growth using Artificial neural network. Eur. J. Sci. Res, 97, 516–526.

Papalexandratou, Z., & Vuyst, D. L. (2011). Assessment of the yeast species composition of cocoa bean fermentations in different cocoa-producing regions using denaturing gradient gel electrophoresis. FEMS Yeast Res, 11, 564–574.

Parapouli, M., Vasileiadis, A., Afendra, A.S., & Hatziloukas, E. (2020). Saccharomyces cerevisiae and its industrial applications. AIMS Microbiol, 6(1), 1–31. doi: 10.3934/ microbiol. 2020001

Peng, W., Zhong, J., Yang, J., Ren, Y., Xu, T., Xiao, S., & Tan, H. (2014). The artificial neural network approach based on uniform design to optimize the fed-batch fermentation condition: Application to the production of iturin A. Microb Cell Fact, 13, 54–54. doi:10.1186 /1475-2859-13-54

Rezende, C.P., Almeida, M.P., Brito, R.B., Nonaka, C.K., & Leite, M.O. (2012). Optimisation and validation of a quantitative and confirmatory LC-MS method for multi-residue analyses of lactam and tetracycline antibiotics in bovine muscle. Food Addit. Contam, 29(4), 541–549.

Rizwan, M., Almansoori, A., & Elkamel, A. (2019). An overview on synthesis and design of microalgal biorefinery configurations by employing super structure-based optimization approach. Energy Syst, 10, 941–966.

Rohlfs, M., & Churchill, A.C.L. (2011). Fungal secondary metabolites as modulators of interactions with insects and other arthropods. Fungal Genet. Biol, 48, 23–34. doi: 10.1016/j.fgb.2010.08.008

Roosendaal, R., Kuipers, E.J, van den Brule, A.J., Peña, A.S., Meuwissen, S.G.M., Walboomers, J.M.M., & Graaff, J.D. (1993). Detection of Helicobacter pylori DNA by PCR in gastrointestinal equipment. Lancet, 341, 900. DOI: https://doi.org/10.1016/0140-6736(93)93112-E

Russo, P., Fares, C., Longo, A., Spano, G., & Capozzi, V. (2017). Lactobacillus plantarum with Broad Antifungal Activity as a Protective Starter Culture for Bread Production. Foods, 6, 110.

Salvetti, E., & O'Toole, P.W. (2017). When regulation challenges innovation: The case of the genus Lactobacillus. Trends Food Sci. Technol, 66, 187–194.

Sattar, S.A., Jacobsen, H., Rahman, H., Cusack, T.M., & Rubino, J.R. (1994). Interruption of rotavirus spread through chemical disinfection. Infect. Control Hosp. Epidemiol, 15, 751-6.

Sattar, S.A., Springthorpe, V.S., & Rochon, M. (1998). A product based on accelerated and stabilized hydrogen peroxide: Evidence for broad-spectrum germicidal activity. Canadian J Infect Control, 23-30.

Senz, M., van Lengerich, B., Bader, J., & Stahl, U. (2015). Control of cell morphology of probiotic *Lactobacillus acidophilus* for enhanced cell stability during industrial processing. Int. J. Food Microbiol, 192, 34–42.

Senz, M., Keil, C., Schmacht, M., Palinski, S., Cämmerer, B., & Hageböck, M. (2019). Influence of Media Heat Sterilization Process on Growth Performance of Representative Strains of the Genus Lactobacillus. Ferment, 5, 20. https://doi.org/10.3390/fermentation5010020

Sewsynker-Sukai, Y., Faloye, F., & Kana, E.B.G. (2017). Artificial neural networks: an efficient tool for modelling and optimization of biofuel production (a mini review). Biotechnol. Biotechnol Equip, 31(2), 221-235. https://doi.org/10.1080/13102818.2016.1269616.

Shang, Z., Li, X.-M., Li, C.-S., & Wang, B.-G. (2012). Diverse secondary metabolites produced by marine-derived fungus Nigrospora sp. MA75 on various culture media. Chem Biodiv, 3, 338–1348. doi: 10.1002/cbdv.201100216.

Sharma, S., Kuila, A., & Sharma, V. (2018). Sterilization Techniques used in Fermentation Processes. Principles and Applications of Fermentation Technology, Book Editor(s):Arindam Kuila,Vinay Sharma, Scrivener Publishing LLC, 45-52. https://doi.org/10.1002/97811194 60381.ch3.

Singh, N., Rai, V., & Tripathi, C. (2012). Production and optimisation of oxytetracycline by a new isolate Streptomyces rimosus using response surface methodology. Med. Chem. Res, 21, 3140–3145. doi: 10.1007/s00044-011-9845-4

Singh, V., Haque, S., Niwas, R., Srivastava, A., Pasupuleti, M., & Tripathi C.K.M. (2017). Strategies for Fermentation Medium Optimization: An In-Depth Review. Front. Microbiol, 7, 2087. doi: 10.3389/fmicb.2016.02087

Singh, V., Khan, M., Khan, S., & Tripathi, C.K. (2009). Optimization of actinomycin V production by Streptomyces triostinicus using artificial neural network and genetic algorithm. Appl. Microbiol. Biotechnol, 82, 379–385. doi: 10.1007/s00253-008-1828-0

Taormina, P.J., & Beuchat, L.R. (1999). Behavior of enterohemorrhagic Escherichia coli O157:H7 on alfalfa sprouts during the sprouting process as influenced by treatments with various chemicals. J. Food Prot, 62, 850-6.

Taylor, Phillip, A., Inaba, Yutak, and Pinto, & Ian J. (2016). "Large-Scale Fermentation of E. coli for the Production of High-Purity Isoprene". Senior Design Reports (CBE), 79. https://repository.upenn.edu/cbe_sdr/79

Tong, L., Bénard, P., Zong, Y., Chahine, R., Liu, K., & Xiao, J. (2021). Artificial neural network based optimization of a six-step two-bed pressure swing adsorption system for hydrogen purification. Ener AI, 5, 100075. https://doi.org/10.1016/j.egyai.2021. 100075.

Turner, F.J. (1983). Hydrogen peroxide and other oxidant disinfectants. In: Block SS, ed. Disinfection, sterilization, and preservation. Philadelphia: Lea & Febiger, 240-50.

Vázquez-Ojeda, M., Segovia-Hernández, J.G., Hernández, S., Hernández-Aguirre, A., & Kiss, A.A. (2013). Design and optimization of an ethanol dehydration process using stochastic methods. Sep. Purif. Technol, 105, 90-97.

Viggiano, A., Salo, O., Ali, H., Szymanski, W., Lankhorst, P., Nygård, Y., Bovenberg, R.A.L., & Driessen A.J.M. (2017). Elucidation of the biosynthetic pathway for the production of the pigment chrysogine by Penicillium chrysogenum. Appl. Environ. Microbiol, 84, e02246-17. doi: 10.1128/AEM.02246-17

Vo, N.D., Oh, D.H., Kang, J.H., Oh, M., & Lee, C.H. (2020). Dynamic-model based artificial neural network for H2 recovery and CO2 capture from hydrogen tail gas. Appl Ener, 273, 115263. https://doi.org/10.1016/j.apenergy.2020.115263

Weber, D.J., & Rutala, W.A. (2004). Disinfection and sterilization of potential bioterrorism agents. In: Rutala WA, (eds.) Disinfection, sterilization and antisepsis: Principles, practices and new research. Washington DC: Association for Professionals in Infection Control and Epidemiology, 86-103.

Wei, X., Yang, H., & Huang, W.T. (2021). A Genetic-Algorithm-Based Optimization Routing for FANETs. Front. Neurorobot, 15, 697624. doi: 10.3389/fnbot.2021.697624

Wong, K.-W., Yap, W.-S., Wong, D.C.-K., Phan, R.C.-W., & Goi, B.-M. (2020). Cryptanalysis of genetic algorithm-based encryption scheme. Multimedia Tools Appl, 79, 25259–25276.

Yadav, M., Sehrawat, N., Sangwan, A., S., Kumar, Beniwal, V., & Singh, A.K. (2013). Artificial Neural Network (ANN): Application in media optimization for industrial microbiology and comparison with response surface methodology (RSM). Adv. Appl. Sci. Res, 4(4), 457-460.

Yahya, A.Z, Hussain, M.A., & Wahab, A.K.A. (2015). Modeling, optimization, and control of microbial electrolysis cells in a fed-batch reactor for production of renewable biohydrogen gas. Int J Energy Res, 39(4), 557–572.

Zhang L., Tang Y., Guo Z. P., Ding Z. & Shi, Y.G.Y. (2011). Improving the ethanol yield by reducing glycerol formation using cofactor regulation in Saccharomyces cerevisiae, Biotechnol. Lett, 33 , 1375 -1380.

Zhang, H., Xie, J., Ge, J., Zhang, Z., & Zong, B. (2019). A hybrid adaptively genetic algorithm for task scheduling problem in the phased array radar. Eur. J. Oper. Res, 272, 868–878.

Zhong, J., Zhang, X., Ren, Y., Yang, J., Tan, H., & Zhou, J. (2014). Optimization of Bacillus subtilis cell growth effecting jiean-peptide production in fed batch fermentation using central composite design. Electron. J. Biotechnol, 17, 132–136. doi: 10.1016/j.ejbt.2014.04.010

Chapter-2

MICROBIAL BIOMASS PRODUCTION FROM BIOLOGICAL PROCESSES

Rajesh Kumar Srivastava[1], Prakash Kumar Sarangi[2], Akhilesh Kumar Singh[3] and Uttam Kumar Sahoo[4]*

[1]Department of Biotechnology, GIT, Gandhi Institute of Technology and Management, Visakhapatnam, India

[2]College of Agriculture, Central Agricultural University, Imphal, Manipur, India.

[3]Department of Biotechnology, School of Life Sciences, Mahatma Gandhi Central University, Motihari, Bihar-845401, India.

[4]Department of Forestry, Mizoram University, Aizawl-796004, India

**Corresponding author Email: rajeshksrivastava73@yahoo.co.in*

Abstract

Microbial biomass synthesis is a very important process via different types of microbial cells system and it is responsible for yield and titer of any metabolites and also others nutrient contents in their biomass synthesis. Microbial biomass is natural and laboratory induced processes and is affected by culture conditions, nutrient availability, biological cell structure pattern and also favors of physical parameters such as temperature, pH, oxygen content levels and its presence /absence or others factors. Nowadays microbial biomass cultivation can help in many regards such as in microalgal biomass and photosynthetic bacterial cells and also yeast cells that can provide many value added products such as protein sources, biofuels and some others bioproducts. Microalgal biomass cultivation can help in environmental remediation and also biodiesel generation that can double benefits from one type of biomass synthesis. For biodiesel synthesis, wet biomass is utilized via sets many steps including uses of organic solvents and suitable catalyst under appropriate conditions. Photosynthetic bacteria can evolve biohydrogen with utilization of sunlight energy. These can be grown in anaerobic conditions in presence of light. Cyanobacteria also utilizes for hydrogen production. Author will explore all impacts of microbial biomass for these products

Keywords: *Microbial biomass, Valuable products, biological process, Parameters, Bioenergy, Cultivation, Nutrients*

2.1. Introduction

Syntheses of different types of microbial biomass are found from natural and laboratory scale growth. And these biomass, now a days utilized for sustainable mode of fuel energy and also others value added product. We can fulfill ever increasing their energy consumption demand for explosive growth of human population and also more developmental activities for today life. Now we are able to develop the high energy conversion approaches/techniques with energy storages facilities/ devices. Now these innovative techniques require for advanced electrode materials and catalysts (Deng et al., 2016)

In current period, researchers have put much effort for laboratory/ scientifically modes biomass cultivation and it can be found more efficient and effective. In this context, we explore the approach for experimentally designed parameters for microbial biomass cultivation and it was implemented at laboratory level to check its efficiency. This approach was discussed as a simple glucose-limited feeding strategy for cultivation of yeast biomass in a bioreactor (i.e., bubble column reactor) (Vieira et al., 2013). This mode of bioreactor was suitable for industrial scale biomass and it was based on a spreadsheet simulator with perfect response to biomass cultivation. In this bioreactor, the utilized cultivated biomass was *Saccharomyces cerevisi*ae strain and this yeast biomass cultivation constraint was strong tendency of these cells species for metabolizing the sugars in anaerobic conditions and it was created due to catabolites repression effects (Akinyemi et al., 2005).

And finally low concentration/ yield of this yeast biomass were found due to on glucose substrates. So this issue was solved due to change/ control in usual strategy for minimization of metabolic tendency and fed-batch process was chosen for this task. This modified process for this yeast biomass was applied with sugar sources feeding plan of incrementally and then they maintained total sugars concentration in fermentation broth that was below to determined value (Frick and Witmann, 2005; Vieira et al., 2013). And this pubished works is discussed for simulator function and simulator was developed to control the molasses feeding rates on the basis of the simple theoretical model and it was taken consideration for nutritional growth needs of yeast cell. Next, input data was considered as theoretical specific growth rate and also initial cell biomass (Frick and Witmann, 2005).

This work results was assayed via use of a commercial baker's yeast strain and molasses as sugars sources for their cultivation in this reactor. This experimental designed found the overall biomass yield on this substrates was 0.33 with a biomass increases upto 6.4 times more biomass. Next, specific growth rate of this yeast was 0.165 h^{-1} with opposite to the predicted value (180 h^{-1}) in the second stage simulation (Ruiz-Martinez et al., 2012; Vieira et al., 2013).There is reports on application of synthesis of photosynthetic bacterial (PSB) biomass that helped in removal of pollutants in water bodies and it has shown its effective to increase its biomass and pigment synthesis. This bacterial species can utilize for wastewater treatment task at different light sources (i.e., red, yellow, blue and white LED light with incandescent lamp uses). This literature discussed its results with effects of different sources of light on PSB biomass yield/ concentration (Kumar et al., 2010).

This literature talked on highest biomass production (2.6 gL^{-1}), chemical oxygen demand (COD~0.5mg biomass/mg. COD removal) and biomass yield (88%) via the effect of red LED light sources. This literature also explored other outcome as shorten hydraulic retention time to 72 h for wastewater treatment with the uses of red LED light for this PSB cultivation. And this modification analysis due to this light source was done by mechanism analysis with high ATP generation due to red light influences and it was compared to others light sources (Papazi et al., 2008). The pigment synthesis in this bacterial biomass also was significantly affected due to influence of different light sources. And yellow color LED was found to effective for highest quantity of pigment biosynthesis but red LED light produced the highest quantity of carotenoid and bacteriochlorophyll ratio. This red light source is both efficient and energy cost at optimal light source (Papazi et al., 2008; Zhou et al., 2015).

This chapter will discuss different types of microbial biomass, cultivation condition, biomass conversion approaches and also some value added products and can see in figure 2.1.

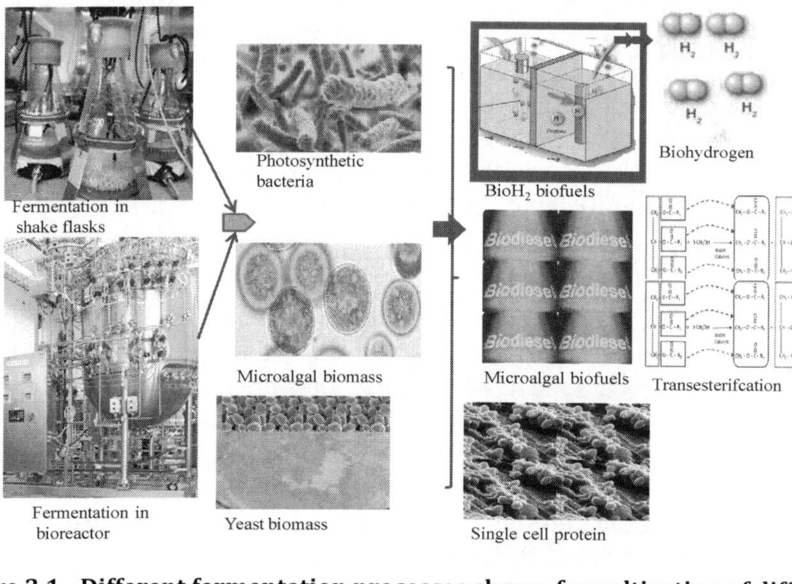

Figure 2.1 - Different fermentation processes shown for cultivation of different types of microbial system biomasses that can later used for different types of value products such as single cell protein, microalgal biodiesel and biohydrogen synthesis

2.2. Microbial Biomass

We are known to various types of microbial biomass and these are synthesized during biological process required for their growth, development and also synthesizing of metabolites and enzymes. These biomasses can synthesize via bacterial, fungal, yeast, blue green algal, macro or microalgae, and also plant cells. And these biomass synthesizing can help in environmental remediation and also help in biofuel development approaches. And our dependency for bioproducts is more from natural processes via involving of microbial biomass catalysis processes. And these can help to synthesize these bioproducts (bio-pigment, single cell protein, fuel energy, different types of organic acid, alcoholic fuels and others products) (Zhou et al., 2015).

2.2.1. Bacterial Biomass

In food industry wastewater, purple-non-sulfur bacteria (PNSB) are found to cultivate via using of anaerobic membrane photo-bioreactor and this bacterial cultivation can help in organic compound removal and biomass biosynthesis. These bacterial cells shown with these characteristics that can be accomplished with an explicit their examination. This bacterial cell found for long performance of the photo-bioreactor fed with real wastewater (Chitapornpan et al., 2013). It has found that with the support of infra-red light transmitting filter, PNSB cell can survive with maintaining in the system under the continual fluctuations of influent wastewater characteristics (Okubo and Hiraishi 2007). This literature research work discussed the average BOD (51%) and COD (58%) efficiencies with moderate values. This bacterial cell showed the better yield of photosynthetic biomass (0.6 g dried solid/g BOD) with crude protein content (0.41g/g dried solid). And denaturing gradient gel electrophoretic analysis (DGGE) and 16 rRNA sequencing approaches provided the confirmation of presence of *Rhodopseudomonas palustris.* This bacterial species can make significant changes in the photo-synthetic bacterial community with in bacterial system (Kaewsuk et al., 2010; Chitapornpan et al., 2013).

Another bacterial biomass from *Bacillus subtilis* strain B7-S strain is discussed as endospore forming and aerobic microorganism and it is presented as model microbial system with gram positive bacteria. This bacterial species shown its capability for bioconversion of ferulic acid into vanillin compound and these properties is confirmed as it basic property. Conversion of ferulic to vanillin is seen in *B. subtilis* B7-S with very effective and best capacity (Ashengroph et al., 2011). This conversion is found under specific temperature (35^0C value), pH (9.0) and shaking speed at 200rpm. And under these conditions, conversion of these product for several, repeated small-scale batch mode experiments was performed with maximum conversion efficiencies (63.3%) after completion of three hours (Ghosh et al., 2007).

Under this bioconversion via this microbial system, vanillin products formation confirmed by spectral data that was achieved from vis, inductively coupled plasma emission spectroscopic (ICP-AES) and FTIR (Fourier transformer infrared spectroscopy). SEM (scanning electron microscopy) and TEM (transmission electron microscopy) analysis techniques have helped to confirm the cell surface of *B. subtilis* microbial cells (Chen et al., 2016). These confirmations of this product and microbial system are playing important role in induction of ferulic acid tolerance. Further it demonstrates the B. subtilis B7-S strain can potential in vanillin production with its uses for ferulic acid conversion process (Ghosh et al., 2007; Chen et al., 2016).

Bacterial subtilis is shown to express the inherited background and it is also found to simple with diverse genetic manipulation in modified system. This microbial system is Gram positive model bacterium with full studies on physiology and metabolism system/ process. Further this bacterial species is shown to exhibit the highly efficient protein secretion system and can show the adaptable metabolism (Tanaka et al., 2017). This microbial system reported to widely use as cell factory for microbial biosynthesis of bio-chemical, enzymes, or antimicrobial material or compounds with more uses in industry, agriculture and medicine field. For this microbial system basic genetic manipulation tool and expression system is discussed as best test/ model organism (Jin et al., 2016).

And also discusses the uses of traditional approaches/ method and also novel engineering system. This microbe can apply for many applications such as synthesis of chemical and enzymes with its advantages. Some reports discussed products formation and recent progress in engineering tasks in *B. subtitlis* strain (Liu et al., 2020). For this microbe system, many applications such as microbial additives, biofilm, spore formation and antimicrobials agents synthesis. Many researchers are working in this area that offers them a better understanding of *B. subtilis* strain and its application in wide range (Tanaka et al., 2017; Liu et al., 2020).

There are many literatures on biosurfactants compound synthesis and its applications. These biosurfactants are produced by many bacterial species biomasses and this is surface-active amphiphilic compounds and showed wide ranges of application. But, production of these biosurfactants compound showed high cost and this cost is severely restricts for their uses. Now it is needed for cheaper rate of production cost of biosurfactant and it can be better chance for uses of indigenously isolated microbial systems that can produce the biosurfacant in high yield and efficient ways (Kumar et al., 2010).

In this context, *Bacillus subtitlis* ANR88 found to exhibit this metabolite formation capability and can grow on different cheap carbon sources such as molasses, milk whey, and extract of potato peels, organic peels, banana peels and bagasse). This microbial strain can use significant amounts/ quantity of total sugars from these cheap carbon raw substrates and was found to produce more quantity cell biomass and also biosurfactant (Plaza et al., 2014). This microbe strain can utilize the minimal media containing glucose and molasses (as sole carbon) and it produce biosurfactant compound (~ 0.207 $g.L^{-1}$ and 0.241 $g.L^{-1}$) concentration respectively. But, milk whey utilization as carbon source this microbial is unable to produce biosurfactant. We discuss the different natures of extracts of the agrowastes such as bagasse and orange peels, these wastes can produce the 0. 127 and 0.09 g.L^{-1} of biosurfactant respectively (Rane et al., 2017).

Literatures have discussed the application of One-variable-at-a-time (OVAT) and are found for helping in optimization of biosurfactant biosynthesis (with produced concentration~0.513 g.L^{-1}) via using of *B. subtilis* ANR88 strain. This biosurfactant concentration was found in media compositions such as molasses (4%), ammonium ferric citrate (0.25%) and pH: 7. Next, researchers used the Plackett-Burman design based statistical method that optimized the fermentation process for increased biosynthesis of biosurfactant (0.746 g.L^{-1}) and this is 3.6 increased improvement in this biosurfactant concentration/ titer on utilized glucose as raw carbon substrate (Gudiña et al., 2015). And next, biosurfactant production from this microbial system (i.e., *B. subtilis* ANR88 strain) was analyzed from the approach/ technique such as Fourier Transform Infra-red spectroscopy. This technique helped in alkyl and peptide groups in this process produced biosurfactant (Plaza et al., 2014; Gudiña et al., 2015).

From this microbial system produced biosurfactant can be found to more effective if it is used in the synthesis of silver and gold nanoparticles that was totally absence in conventional chemical agents. For our interest point view, these nanoparticles production is nearly uniform in their sizes and shapes with spherical shape silver (4-18nm) and hexagonal gold nanoparticles (40-60nm) as confirmed by TEM technique (Plaza et al., 2014; Rane et al., 2017). Utilization of *Bacillus thuringiensis* is found this microbial based insecticides and is found to get this bioactive component in cost-effective process with need its for development at large scale. It needs for cultivation of this bacterial strain in rich medium that can help to obtain the highest yield of spore-crystal complexes (Lopezy-Lopez and de la Torre, 2005). This bacterium in this cultivation medium (with glucose concentration~ 50-90 g.L^{-1}) can show the results as much lower crystal protein, lower concentration of bacterial spore and lower toxicity after testing against with larva of *Spodoptera littoralis* and *Anagasta kuechniella* (Alotaibi et al., 2008).

And glucose concentration at 20g.L^{-1} media composition can able protein 7.1x10^{11} spore/ml and 3.4 g.L^{-}1 of crystal protein is achieved with LC50 of 40.1 and 50.2 mg/kg meal against *S littoralis* and *A. kuehniella* respectively. Finally, greater than 21% of consumed glucose substrates were shown to diverted into synthesis of by-products at the expenses of spore-crystal protein mixtures (Rowe et al., 1987). And 78.3% of consumed glucose was found for its conversion into spore and crystal protein. And formed byproducts are acetic acid and β- hydroxybutyrate during active growth phase. Others are succinic and glutamic acid, synthesized during the active phase of toxins (Rowe et al., 1987; Alotaibi et al., 2008).

2.2.2. Fungal Biomasses

Plant lignocelluloses are found to bioconvert into second generation biofuel and biochemicals. These conversions are reported by fungal species and plays essential role in carbon cycling in terrestrial ecosystem. But, lignocellusoic material is contained recalcitrant aromatic polymers lignin as a major obstacle in biofuels and biochemicals biosynthesis process (Mäkelä et al., 2015). So, it needs to go for microbial degradation of lignin with receiving a great deal in many researchers and industrial people interest. Any species of fungi are known as main degrader of plant biomasses especially this property is seen in basidomycete white rot fungi. This fungal species is found to convert these plant aromatics into simpler one (Lubbers et al., 2019).

This fungal species is shown its ability for degradation of this lignin component of plant. And released aromatic monomers from this lignin component of plant biomass are toxic in nature for most fungi at low level. And from this microbial system, conversions of these compounds are found to less toxic metabolite in nature (Kowalczyk et al., 2019). These metabolites could be essential for fungi. Number of researchers have studied the released aromatic compounds from lignin of plant biomass and now showed more attention for metabolic pathways that can convert these resultant aromatic monomers. Further, details are given for these plant aromatic compounds that are released from plant biomass via fungal species degradation processes (Kowalczyk et al., 2019; Mäkelä et al., 2015).

In fungal biomass synthesis, aromatic compound such as ferulic acid andp-coumaric acid is released by *Aspergillus niger* and these compounds are precursor for vanillin and p-vinyl phenol respectively. Nowadays around compounds is found as very important molecules and widely applied in many industries and earlier time it was mainly synthesized from non-renewable sources/ feedstock. In current period, there are many sources of renewable biomasses or waste matters available such as plant biomass and it can provide interesting alternative for production of these aromatic compounds (Lubbers et al., 2021).

Further, degradation of hydroxycinnamic acids i.e., caffeic acid, ferulic acid and p-coumaric acid is found from many species of fungi and *A. niger* is known to contain multiple metabolic pathways that can help in degradation hydroxycinnamic acids. But no specific genes were observed for these hydroxycinnamic acids metabolic pathways. Many studies were done on genes on several pathways via using whole-genomic transcriptomic data of *A. niger* strain (Santamaría et al., 2018). And this fungal species was cultivated on different hydroxycinnamic acids concentration. And the genes in involved in the CoA-dependent β- oxidative pathways in these fungi are reported and this pathway was involved in degradation of fatty acids but for the hydroxycinnamic acids. But in plants, this pathway was shown to degrade the hydroxycinnamic acids (Gabriel et al., 2014).

For this pathway, identified genes coding enzymes are hydroxycinnamate–CoA synthase (hcs A), multi-functional β- oxidation hydratase/ dehydrogenase (foxA), 3-ketoacyl AcA thiolase (katA) and fours thioesterases (the A-D) of *A. niger* (Gabriel et al., 2014; Santamaría et al., 2018). All these genes were highly induced and expressed by all types of hydroxycinnamic acids. All genes deleted in mutant strain can confirm for its indeed to involve in degradation of several hydroxycinnamic acids. It is known for foxA and the B genes that involved in degradation of fatty acid but HcsA, FoxA and KatA genes are found to contain a peroxisomal targeting signal that are localized in peroxisomes cell organelles (Santamaría et al., 2018; Lubbers et al., 2021).

For fungal biomass, a new post-treatment approach is applied after fermentation process and this approach is found to utilize for pigment production as an additional valuable metabolites. This post-treatment approach is found to incubate at 95% relative humidity with studies of others factor such as harvesting time, light and temperature value. In this regard, fungus *Neurospora intermedia* is cultivated on ethanol plant residues that produced ethanol (4 $g.L^{-1}$) and fungal biomass (5 $g.L^{-1}$) (Gmoser et al., 2019). And this fungal biomass with edible filamentous nature is found with pigment production capacity. This fungal harvesting can show pale biomass after 48h submerged cultivation and it was compared to 14h or 72h period of fermentation is found for increased pigmentation in the post-treatment step with 35% and 48% respectively. And this fungal biomass can produce 1.4mg/g dry fungal biomass and this value can obtain from washed biomass treated in light at 35^0C (Ferreira et al., 2014).

Further major impact on pigmentation was found from washed biomass. Further, post-treated biomass is found to 50% w/w crude protein. From this literature, it has shown the post-treatment approach can able to successfully add to pigments to pre-obtained biomass. This pigmented fungal biomass can find the animal feed application for domestic animals (Khiabani, et al., 2011; Gmoser et al., 2019). Dry mill is utilized for whole grain processing and uses of ethanol process via utilizing this waste can produce ethanol and animal feed as combined benefits. And wastewater generation after the distillation and separation of solid materials is known as thin stillage. And utilization of thin stillage can found for production of 3.5 $g.L^{-1}$ ethanols and 5.0$g.L^{-1}$ of biomass during batch fermentation (Ferreira et al., 2015).

This mode of fermentation was completed via uses of fugal *Neurospora intermedia* in a 2 meters high airlift reactors and bubble column. This batch modes cultivated fungal biomass is reported to 50% w/w protein, 12% of w/w lipid. Further, this biomass is also rich in essential amino acids and omega-3 and -6 fatty acids (Aranaz et al., 2009). And in continuous mode of fermentation, dilution rate was found to 0.2 h^{-1} and can be applied without cell washout in the bubble column at 0.5vvm. And 0.1h^{-1} of dilution rate is shown to produce high ethanol (around 5$g.L^{-1}$) titer and biomass (4$g.L^{-1}$) and these biomass are reported to contain nearly 50% protein (Lennartsson et al., 2014).

And this fungal biomass can able to assimilate saccharides in the liquid fraction and sugar backbones (i.e. xylan and arabinan) in the solid fraction. And the inclusion of the current process can find to potentially lead for production of 11000 m^3 of ethanol (5.5% improvement via compared to normal industrial process. And this value can be found around 6300 tons of high quality biomass for animals feed at a typical facility for production of 2 x 10^5 m^3 ethanol/year (Lennartsson et al., 2014; Ferreira et al., 2015).

Fungal metabolites with enhanced biomass are found from long periods of time with common history with human and animal nutrition and also health inducing nutrients. In this regard, Macro and filamentous fungal species and these species are good sources of proteins, lipids, vitamins, pigments, colorant, bioactive compounds, antibiotics, oligosaccharides, minerals and pharmaceuticals etc. (Aguilar-Pérez et al., 2020). And its good example is industrially important enzymes and microbial biomass proteins that are biosynthesized from many fungal species for more than 50 years. And there are some reports on start-ups conversion of by-products and side streams are found to rich sources of carbohydrates and can be converted into a protein-rich fungal biomass (Venil et al., 2020).

Now, this fungal biomass is processed into vegan meat substitute that can be utilized in food as its application and last few years, people have work lots and there is significant rise in international level literatures that discussed the lipid production from microbial fungal sources. More common is single cell oils (SCOs) are biosynthesized from oleaginous microbial system including oleaginous fungi (i.e. zygomycete species especially *Cunninghamella echinulata* and *Mortierella isabellina*) (Dufossé et al., 2021). As we known about species of fungi that are potential sources of polyunsaturated fatty acids (PUFAs) as this nature of microbial system with capability to accumulate large quantity of high valued PUFAs. And PUFAs can be gamma-linolenic acid (GLA) and arachidonic acid (ARA) (Dzurendova et al., 2020).

There are many reports published in many journals that discussed the importance of fungi and fungal metabolites with properties of improving human and animal nutrition and health (Takahashi et al., 2020). And some application of these fungi and its metabolites are explored with its example such as bioactive fatty acids, pigments, polysaccharides, alkaloids, terpenoids. These compounds found to provide the great potential in human and animal nutrition and health (Takahashi et al., 2020; Dufossé et al., 2021). There are many plant diseases, caused by phytopathogenic fungal and it can lead to heavy/ huge losses in agricultural crops and its sections/ fields. And these diseases can continuously threat to the global food security. In this regards, chemical-based fungicides, can be contributed in significant role in securing more quantity of crop production (Xu et al., 2021).

Now, these efforts of application of fungicides in continuous can lead to increased chemical resistance and potential risks to human health and environment. And there is very urgent need for searching for new bioactive natural products and it can utilize in development of new biopesticides (Zhao et al., 2020). Now, fungal endophytes microbial system can reside in the fresh tissues of living plant and it counted as untrapped sources of novel natural products for exploitation in agriculture and or medicine. And chemical examination of endophytic fungal species is reported with yields of various types of antifungal natural products that can apply in potential ways for development of biopesticides (Zhu et al., 2017).

There are nearly 132 antifungal metabolites are isolated from different fungal endophytes from last two decades. And researchers have given emphasis on the unique chemical diversity of these metabolic products together with their relevant antifungal properties. Some main metabolites such as grisofulvin, trichothecene and synthetic derivatives that posses high potential as candidates of new natural fungicides, shown / reported (Zhao et al., 2020; Xu et al., 2021).

2.2.3. Yeast Biomass

Biosynthesis of biofuels from utilization of lignocellulosic biomass or its residues is explored and from this feedstock, biofuel synthesis is unsolved and complicated challenges in industrial biotechnology. Now, many types of microbial system including yeast cells are also utilized with solving many challenges that raised from host organism with their uses for conversion these wastes into biofuels (Jin and Cate, 2017). Now these issues/ challenges can be solved by applying genetic manipulation with robust industrial phenotypes characteristics. In this context, *Saccharomyces cerevisiae* strain can utilize for premier host for metabolic engineering of biofuel generation pathways. Except it, there is others many genetic, systems and synthetic biology tools that can also apply for modification of these bioprocesses (Ko and Lee, 2018).

Now, there are numerous engineering approaches available that can expand the substrates ranges for its consumption and also various types of products formation from engineered *S. cerevisiae.* And wild types of others yeast species is generally lack these modification tools and are unable to harbor superior phenotype or characters. Recently, modified yeast stain can be exploited to manage in harsh process and it needed in lignocellulosic based biofuel biosynthesis (Abdel-Mawgoud et al., 2018).

In lignocellulosic feedstock utilization, some harsh conditions are thermotolerant, resistance to toxic compounds (that are generated during plant biomass deconstruction/ degradation) and also wide range of carbon consumption capability that are needed to adapt. Contrast to *S. cerevisiae* strain, oleaginous yeast such as *Yarrowia lipotytica* strain, is found to capable of high titers of lipids synthesis and accumulation and it can be utilizing via developing into advanced form with application of tools such as metabolic manipulation (Jin and Cate, 2017; Abdel-Mawgoud et al., 2018). In present time, current transportation need of fuel is derived from petroleum/ its byproducts and these fuels can develop from different microbial systems and in this regard, especially oleaginous yeast are naturally found and also evolved that can show the high flux pathways for fatty acids (in form of neutral lipid) synthesis and accumulation (Spagnuolo et al., 2019).

These fatty acids can be easily converted into a variety of drop-in fuels. Many reports have described the recent advanced in the uses of the four most popular oleaginous yeasts species and are reported for making lipids and others potentials fuels. These yeast species are *Yarrowia lipotytica, Lipomyces starkeyi, Rhodosporidium toruloides* and *Cutaneotrichosporon oleaginuous* and these have been explored in various tasks, focusing on important natural complex phenotype of each yeast species, and next is development and utilization of metabolic engineering tools for each species (Sreeharsha and Mohan, 2020). Next focus was demonstration of metabolic engineering impact in yeast species. All these efforts in improvement of these yeasts capabilities can be found most promising in the near and long term for bioenergy biosynthesis (Chaturvedi et al., 2018; Sreeharsha and Mohan, 2020).

In regard to fuel energy synthesis, oils and oleochemicals production is found by microbial cells and it can best and attractive alternative to petroleum. And food based oils for the biosynthesis of transport oils and oleochemical products. Now an emerging candidate for industrial single cell oils oil production is found with help of oleaginous yeast *Lipomyces starkeyi*. This yeast has shown its capability for accumulation of high quantity of lipids (concentrations greater than 60% of dry cell weight) (McNeil and Stuart, 2018).

From the perspective of industrial biotechnology, *L. starkeyi* is one of best candidate for single cell oil and oleochemical biosynthesis and these can utilized for wide variety of carbon and nitrogen sources as a good feedstock. *L. starkeyi* strain is found to use for high quantity of lipid biosynthesis and accumulation from utilization of hexose and pentose sugar that can be generated from cellulosic hydrolysates, crude glycerol and sewage sludge (Yang et al., 2014). This yeast strain can produce the glucanhydrolases and this sugar can show a variety of industrial application and also played potential role for bioremediation. This has shown excellent properties for biotechnology application and can adapt to best candidate (Xavier et al., 2017). Maximum hindrance can be solved by applying the genetic manipulation methods that can help in engineering of this strain. *L. starkeyi* strain has shown more industrial potential candidates for biosynthesis of lipids, oleochemicals and other biochemicals that showed many industrial applications (Xavier et al., 2017; McNeil and Stuart, 2018).

There are many advanced and recent techniques in bioprocesses are found that can show the significant ways of conversion of lignocellulosic biomass into biofuels development. And these conversion demand is found nearly 90% energy upgrading of cellulosic sugars into generation of renewable nature of drop-in biofuels precursors. The effective hydrogen to carbon ($H_{eff}/C{\sim}2$) can be used for reflect the oxygen level in a potential feedstock (Pawar et al., 2019) and figure 2.2 shows the value added products including biofuels.

And chemical methods for this regard is failed to achieve this value without substantial loss of carbon. And oleaginous biological system can propose a greener upgradation route for production of oils from sugars with 30% theoretical yields (Kruger et al., 2018). And these oleaginous systems can't compete with commercial volumes of vegetables oils in terms of overall oil yields and productivities (Nogué et al., 2018). There is many significant challenges in the commercial exploitation of these microbial oils and it can lie inefficient recovery of the production of oils. Utilization of highly selective oils capturing agent can allow a concomitant microbial oil production and also in situ oil recovery process (Nogué et al., 2018; Pawar et al., 2019).

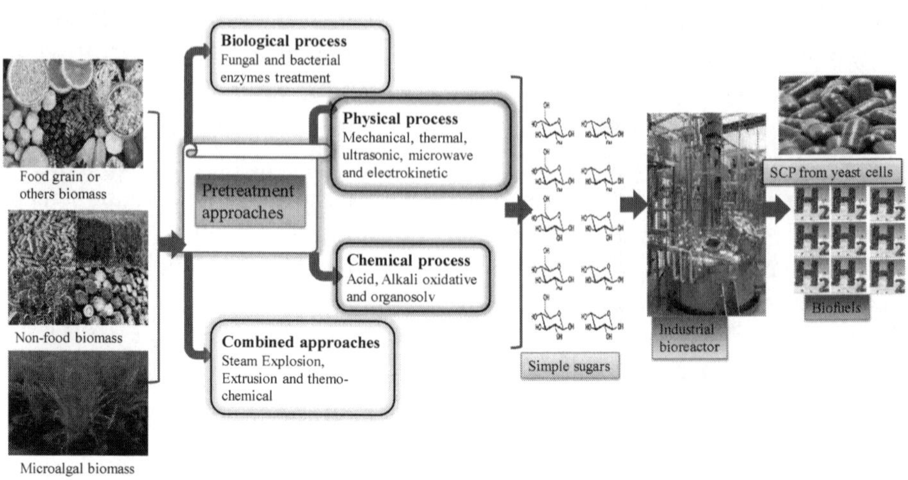

Figure 2.2 - Different sources of microbial biomass utilization for value added products synthesis including sustainable fuel sources for transportation.

Other approaches for estimation biomass of microbial system are reported via including the ergosterol estimation and this is essential components of fungal cell membrane and it can use for determination of fluidity, permeability and active of cells with reports as membrane assisted protein. Ergosterol synthesis in fungal biomass is complex, highly energy-consuming pathways that can involve to participation of many enzymes. And deficiencies in sterol biosynthesis can pleiotropic defects with limitation in cellular proliferation and adaptation to stress. And fungal ergosterol levels are found to tightly controlled process by the bioavailability of particular metabolites (such as sterols, oxygen and iron) and environmental conditions (Jordá and Puig, 2020).

And regulation of ergosterol synthesis is achieved by overlapping mechanism that include transcriptions expression, feed inhibitions of enzymes and changes in their subcellular localization. *Saccharomyces cerevisiae* is known as budding yeast and shown to contain sterol regulatory element (SRF) binding protein Upc2 and Ecm22, heme-binding protein (Hap1) and repressor factors Rox1 and Mot3 and these genes are found to coordinate esrgosterol biosynthesis (ERG) genes expression (Tian et al., 2018). There are some reports that summarize the sterol biosynthesis, transport and detoxification systems of *S. cerevisiae* and it is found to adaptive response to sterol depletion, low oxygen, hyperosmotic stress and iron deficiency. There are still large numbers of ERG and the crosstalk between different environmental signals and pathways available (Romero et al., 2019).

Many aspects of ergosterol regulation are not still known and study of sterol metabolism and regulatory mechanisms is highly relevant due to its wide application in antifungal treatment and also food/ pharmaceutical industries (Romero et al., 2019; Jordá and Puig, 2020) and table 2.1 discusses the some value added products from microbial biomasses.

Table 2.1- Different type of biomasses with their value-added products or nutrients

Biomasses	*Value added products/ nutrients*	*References*
Cultivation of yeast biomass in a bioreactor	Metabolizing the sugars in anaerobic conditions with low concentration/ yield due to catabolites repression effects	Akinyemi et al., 2005
Commercial baker's yeast strain with molasses as sugars sources in this reactor	Overall biomass yield on this substrates was 0.33 with a biomass increases upto 6.4 times more biomass	Ruiz-Martinez et al., 2012;
Synthesis of photosynthetic bacterial (PSB) biomass	Highest biomass production (2.6gL-1), chemical oxygen demand (COD~0.5mgbiomas/mg. COD removal)) and biomass yield (88%) via the effect of red LED light sources	(Papazi et al., 2008
Infra-red light transmitting filter, purple-non -sulfur bacteria (PNSB) cell can survive with maintaining in the system	Better yield of photosynthetic biomass (0.6 g dried solid/g BOD) with crude protein content (0.41g/g dried solid)	Kaewsuk et al., 2010; Chitap-ornpan et al., 2013
Bacterial biomass from Bacillus subtilis strain B7-S	Its capability for bioconversion of ferulic acid into vanillin compound	Ashenroph et al., 2011
Bacterial subtilis studied on physiology and metabolism system/ process	Exhibit the highly efficient protein secretion system and can show the adaptable metabolism	Tanaka et al., 2017
Engineering tasks in B. subtitlis strain discussed products formation	Showed many applications such as microbial additives, biofilm, spore formation and antimicrobials agents synthesis	Liu et al., 2020
Bacillus subtitlis ANR88 found to exhibit this metabolite formation capability	It uses significant amounts/ quantity of total sugars from these cheap carbon raw substrates and produces more quantity cell biomass /also biosurfactant	Plaza et al., 2014
Application of One-variable-at-a-time (OVAT) for biomass	Helping in optimization of biosurfactant biosynthesis (with produced concentration~0.513 gL-1) via using of B. subtilis ANR88 strain	Gudiña et al., 2015
Utilization of Bacillus thuringiensis is found	Used for microbial based insecticides and is found to get this bioactive component in cost-effective process	Lopezy-Lopez and de la Torre, 2005
Conversions reported by fungal species with essential role in carbon cycling in terrestrial ecosystem	Lignocellusoic material \ contained recalcitrant aromatic polymers lignin as a major obstacle in biofuels and biochemicals biosynthesis process	Mäkelä et al., 2015

2.3. Microbial Biomass Estimation Method

Microbial biomass estimation is done by using chloroform fumigating - extraction method and in this approach microbial cells are first killed by exposing fresh soil to ethanol-free chloroform for a 24 h or certain period of time. It can help to extract the carbon, released from lysed microbial cells system with a salt solution (Sullivan et al., 2009). Then, analysis of net increase in soluble carbon content, multiplied by a conversion factor that can help in determination of total microbial biomass. Others common methods are available for determination of microbial biomasses and it can utilize the use of fatty acid synthesis profiles, substrate induced respiration (SIR) or biochemical indicators such as measurement of ATP biosynthesis or fungal ergosterol synthesis by microbial biomass system (Sullivan et al., 2009; Hu et al., 2017)

Microbial biomass examination is done by using very common techniques, used for soil microbial systems in soil microbiology such as availability of DNA sequencing approach. In this regard, several studies is supported to evaluating the influences of harvesting of soil microbial system, employed for microbial biomass measurement tasks with conflicting results. Now, this biomass measurement study is done using a chronosequene approach to examine the microbial community shifts after clear cutting. In literature it is found for total microbial biomass estimation without differing in years following to harvest process (Hynes and Germida, 2013). Other studies are done without loss in microbial biomass from two years after harvesting process. And in contrast to several other studies, there is found a negative and immediate impact of harvest on microbial biomass. And measurement of microbial biomass can provide information about microbial abundance but it can't provide the information on active form that are present in specific locations. The low resolution of any techniques can explain the inconsistent results from similar studies (Holden and Treseder, 2013; Xiong et al., 2011).

2.4. Conclusion

This chapter is mainly focused on different types of biomasses that are formed different biological processes. Now these biomasses are found to provide the best yield and concentration based on different parameters of biological processes such as different modes of fermentation and biotransformation. Now these biomasses are found best sources of various type of nutrients and also raw carbon sources. Now, different types of value added products such as biofuels, pigment, organic acid, alcoholic fuels, and others metabolites and seen in table. As we seen the important media components and its concentration that affect the metabolites and biomass synthesis. This chapter has explored the various types of media composition, fermentation modes and processes parameter. Genetic tool and pathways showed for metabolites improvement and control biomass production. As we know that some biomass are good sources of health caring nutrients that can help in prevention and treatment of discussing with natural way of caring human health.

2.5. Abbreviations

ARA: Arachidonic acid; **ATP**: Adenosine triphosphate; **BOD**: Biological oxygen demand; **COD**: Chemical oxygen demand; **DGGE**: Denaturing gradient gel electrophoretic analysis; **ERG**: Esrgosterol; **foxA:** Functional β- oxidation hydratase/ dehydrogenase; **FTIR**: Fourier transformer infrared spectroscopy; **GLA**: Gamma-linolenic acid; **Hap1**: Heme-binding protein; **hcs**: Hydroxycinnamate–CoA synthase; **ICP-AES**: Inductively coupled plasma emission spectroscopic; **katA**: 3-ketoacyl AcA thiolase; **LC50**: Lethal Concentration fifty; **LED**: Light emitting diode; **OVAT**: One-variable-at-a-time; **PNSB**: Purple-non-sulfur bacteria; **PSB**: Photosynthetic bacterial; **PUFAs**: Potential sources of polyunsaturated fatty acids; **SCOs**: Single cell oils; **SEM**: Scanning electron microscopy; **SRF**: Sterol regulatory element; **TEM**: Transmission electron microscopy;

Template

Chatterjee, C., Pong, F., & Sen, A. (2015). Chemical conversion pathways for carbohydrates. Green Chem, 17, 40–71.

References

Abdel-Mawgoud, A.M., Markham, K.A., Palmer, C.M., & Liu N. (2018). Stephanopoulos G, Alper HS., Metabolic engineering in the host Yarrowia lipolytica. Metab Eng, 50, 192-208. doi: 10.1016/j.ymben.2018.07.016.

Aguilar-Pérez, M.M., Torres-Mendoza, D., Vásquez, R., Rios, N., & Cubilla-Rios, L. (2020). Exploring the Antibacterial Activity of *Pestalotiopsis* spp. under Different Culture Conditions and Their Chemical Diversity Using LC–ESI–Q–TOF–MS. J. Fungi, 6, 140.

Akinyemi, O.P., Betiku, E., & Solomon, B.O. (2005). Substrate Channelling and Energetics of Saccharomyces cerevisiae DSM 2155 Grown on Glucose in Fed-Batch Fermentation Process. Afr J Biotechnol, 2, 96–103.

Alotaibi, S.G.A., Narmen Y.A., & Saleh W.D. (2008). Bioinsecticide production by the bacterium Bacillus thuringiensis. 1. Pattern of cell growth, toxin production and by-product synthesis. Arch Agron. Soil Sci, 54, 387-394. https://doi.org/10.1080/036 50340802158201

Aranaz, I., Mengibar, M., Harris, R., Panos, I., Miralles, B., Acosta, N., Galed, G., & Heras, A. (2009). Functional characterization of chitin and chitosan. Curr. Chem. Biol, 2, 3, 203–230. DOI : 10.2174/2212796810903020203

Ashengroph, M., Nahvi, I., Zarkesh-Esfahani, H., & Momenbeik, F. (2011). Candida galli strain PGO6: a novel isolated yeast strain capable of transformation of isoeugenol into vanillin and vanillic acid. Curr Microbiol, 62, 990–998. doi: 10.1007/s00284-010-9815-y,

Chaturvedi, S., Bhattacharya, A., & Khare, S.K. (2018). Trends in Oil Production from Oleaginous Yeast Using Biomass: Biotechnological Potential and

Constraints. Appl Biochem Microbiol, 54, 361–369. https://doi.org/10.1134/S000368381804004X

Chen, P., Yan, L., Wu, Z. Li S, Bai Z, Yan X, Wang N, Liang N., & Li, H. (2016). A microbial transformation using *Bacillus subtilis* B7-S to produce natural vanillin from ferulic acid. Sci Rep, 6, 20400. https://doi.org/10.1038/srep20400

Chitapornpan, S., Chiemchaisri, C., Chiemchaisri, W., Honda, R., & Yamamoto, K. (2013). Organic carbon recovery and photosynthetic bacteria population in an anaerobic membrane photo-bioreactor treating food processing wastewater. Bioresour Technol, 141, 65-74. https://doi.org/10.1016/j.biortech.2013.02.048

Deng, J., Lia M., & Wang, Y. 2016. Biomass-derived carbon: synthesis and applications in energy storage and conversion. Green Chem, 18, 4824-4854. https://doi.org/10.1039/ C6GC0 1172A

Dufossé, L., Fouillaud, M., & Caro, Y. (2021). Fungi and Fungal Metabolites for the Improvement of Human and Animal Nutrition and Health. Journal of Fungi, MDPI, 7 (4). ff10.3390/jof7040274ff.ffhal-03197753f

Dzurendova, S., Zimmermann, B., Tafintseva, V., Kohler, A., Horn, S.J., & Shapaval, V. (2020). Metal and Phosphate Ions Show Remarkable Influence on the Biomass Production and Lipid Accumulation in Oleaginous Mucor circinelloides. J. Fungi, 6, 260.

Ferreira, J.A., Lennartsson, P.R., & Taherzadeh, M.J. (2014). Production of ethanol and biomass from thin stillage using food-grade Zygomycetes and Ascomycetes filamentous fungi. Energ, 7(6), 3872–3885.

Ferreira, J.A., Lennartsson, P.R., & Taherzadeh, M.J. (2015). Production of ethanol and biomass from thinstillage by *Neurospora intermedia*: A pilotstudy for process diversification Eng. Life Sci, 15, 751–759. DOI: 10.1002/elsc.201400213

Frick, O., & Witmann, C. (2005). Characterization of the metabolic shift between oxidative and fermentative growth in *Saccharomyces cerevisiae* by comparative 13C flux analysis. Microb. Cell Factor, 4, 30.

Gabriel, F., Accoceberry, I., Bessoule, J.-J., Salin, B., Lucas-Gue´rin, M., Manon, S., Dementhon, K, & Noe¨l, T. (2014). A Fox2- Dependent Fatty Acid ß-Oxidation Pathway Coexists Both in Peroxisomes and Mitochondria of the Ascomycete Yeast *Candida lusitaniae*. PLoS ONE 9(12), e114531. doi:10.1371/journal.pone.0114531

Ghosh, S., Sachan, A., Sen, S. K., & Mitra, A. (2007). Microbial transformation of ferulic acid to vanillic acid by *Streptomyces sannanensis* MTCC 6637. J Ind Microbiol Biotechnol, 34, 131–138.

Gmoser, R., Ferreira, J.A., Taherzadeh, M.J., Patrik, R., & Lennartsson, S. (2019). Post-treatment of Fungal Biomass to Enhance Pigment Production. Appl Biochem Biotechnol, 189, 160–174. https://doi.org/10.1007/s12010-019-02961-y

Gudiña, E.J., Rodrigues, A.I., Alves, E., Domingues, M.R., Teixeira, J.A., & Rodrigues, L.R. (2015). Bioconversion of agro-industrial by-products in rhamnolipids toward applications in enhanced oil recovery and bioremediation. Bioresour. Technol, 177,87–93.doi:10.1016/j. biortech. 2014.11.069

Holden, S.R., & Treseder, K.K. (2013). A meta-analysis of soil microbial biomass responses to forest disturbances. Front Microbiol, 4, 163. doi: 10.3389/fmicb.2013.00163

Hu, Z., He, B., Ma, L., Sun, Y., Niu, Y., & Zeng, B. (2017). Recent Advances in Ergosterol Biosynthesis and Regulation Mechanisms in *Saccharomyces cerevisiae*. Indian J Microbiol, 57(3), 270–277. doi: 10.1007/s12088-017-0657-1.

Hynes, H.M., & Germida, J.J. (2013). Impact of clear cutting on soil microbial communities and bioavailable nutrients in the LFH and Ae horizons of Boreal Plain forest soils. Forest Ecol Manag, 306, 88-95. DOI:10.1016/j.foreco.2013.06.006

Jin Y-S., & Cate J.H. (2017). Metabolic engineering of yeast for lignocellulosic biofuel production. Curr Opin Chem Biol, 41:99-106. doi: 10.1016/j.cbpa.2017.10.025.

Jin, P., Kang, Z., Yuan, P., Du, G., & Chen, J. (2016). Production of specific-molecular-weight hyaluronan by metabolically engineered Bacillus subtilis 168. Metab Eng, 35, 21–30.

Jordá, T., & Puig, S. (2020). Regulation of Ergosterol Biosynthesis in Saccharomyces cerevisiae. Genes (Basel), 11(7), 795. doi: 10.3390/genes11070795

Kaewsuk, J., Thorasampan, W., Thanuttamavong, M., & Seo, G.T. (2010). Kinetic development and evaluation of membrane sequencing batch reactor (MSBR) with mixed cultures photosynthetic bacteria for dairy wastewater treatment. J. Environ. Manage, 91, 1161–1168.

Khiabani, M.S., Esfahani, H.Z., Azizi, M.H., & Sahari, M.A. (2011). Effective factors on stimulate and stability of synthesised carotenoid by *Neurospora intermedia*. Nutr. Food Sci, 41(2), 89–95.

Ko, J.K., & Lee, S.M. (2018). Advances in cellulosic conversion to fuels: engineering yeasts for cellulosic bioethanol and biodiesel production. Curr Opin Biotechnol, 50, 72-80. doi: 10.1016/j.copbio.2017.11.007.

Kowalczyk, J.E., Peng, M., Pawlowski, M., Lipzen, A., Ng, V., Singan, V., Wang, M., Grigoriev, I.V, & Mäkelä, M.R. (2019). The White-Rot Basidiomycete Dichomitus squalens Shows Highly Specific Transcriptional Response to Lignocellulose-Related Aromatic Compounds. Front. Bioeng. Biotechnol, 7, 229. doi: 10.3389/fbioe.2019.00229

Kruger, J., Cleveland, N., Yeap, R., Dong, T., Ramirez, K., Nagle, N., Lowell, A., Beckham, G., McMillan, J., & Biddy, M. (2018). Recovery of fuel-precursor lipids from oleaginous yeast. ACS Sustain Chem Eng, 6(3), 2921–31. https://doi.org/10.1021/acssuschemeng.7b01874.

Kumar, C. G., Mamidyala, S. K., Das, B., Sridhar, B., Devi, G. S., & Karuna, M. S. L. (2010). Synthesis of biosurfactant-based silver nanoparticles with purified rhamnolipids isolated from Pseudomonas aeruginosa BS-161R. J. Microbiol. Biotechnol, 20, 1061–1068. doi: 10.4014/jmb.1001.01018

Kumar, M.S., Miao, Z.H.H., & Wyatt, S.K. (2010). Influence of nutrient loads, feeding frequency and inoculum source on growth of Chlorella vulgaris in digested piggery effluent culture medium. Bioresour. Technol, 101, 6012-6018.

Lennartsson, P.R., Erlandsson, P., & Taherzadeh, M.J. (2014). Integra-tion of the first and second generation bioethanol processes and the importance of by-products. Bioresour Technol, 165, 3–8.

Liu, Y., Fang, C., Fang H., & Zhang, D. (2020). *Bacillus subtilis*: a universal cell factory for industry, agriculture, biomaterials and medicine. Microb Cell Fact, 19, 173. https://doi.org /10.1186/s12934-020-01436-8

Lopezy-Lopez, V.E, & de la Torre, M. (2005). Redirection of metabolism during nutrient feeding in fed-batch cultures of B. thuringiensis. Appl Microbiol Biotechnol, 67(2), 254–600.

Lubbers, R.J.M, Dilokpimol, A., Visser, J., Mäkelä, M.R., Hildén, K.S., & de Vries, R.P. (2019). A comparison between the homocyclic aromatic metabolic pathways from plant-derived compounds by bacteria and fungi. Biotechnol Adv, 15; 37(7), 107396. doi:10.1016/j.biotechadv. 2019.05.002

Lubbers, R.J.M., Dilokpimol, A., Visser, J., & de Vries, R.P. (2021). Aspergillus niger uses the peroxisomal CoA-dependent β-oxidative genes to degrade the hydroxycinnamic acids caffeic acid, ferulic acid, and p-coumaric acid. Appl Microbiol Biotechnol, 105(10), 4199-4211. doi: 10.1007/s00253-021-11311-0,

Mäkelä, M.R., Marinović, M., Nousiainen, P., Liwanag, A.J.M., Benoit, I., Sipilä, J., Hatakka, A., de Vries, R.P., & Hildén, K.S. (2015). Aromatic metabolism of filamentous fungi in relation to the presence of aromatic compounds in plant biomass. Adv Appl Microbiol, 91:63-137. doi: 10.1016/bs.aambs.2014.12.001. DOI: 10.1016/bs.aambs.2014.12.001

McNeil, B.A., & Stuart, D.T. (2018). Lipomyces starkeyi: an emerging cell factory for production of lipids, oleochemicals and biotechnology applications. World J Microbiol Biotechnol, 34, 147. https://doi.org/10.1007/s11274-018-2532-6

Nogué, V.S., Black, B., Kruger, J., Singer, C., Ramirez, K., Reed, M., Cleveland, N., Singer, E., Yi, X., Yeap, R., & Linger, J. (2018). Beckham G. Integrated diesel production from lignocellulosic sugars via oleaginous yeast. Green Chem, 20(18):4349–65. https://doi.org /10.1039/C8GC01905C

Okubo, Y., & Hiraishi, A. (2007). Population dynamics and acetate utilization kinetics of two different species of phototrophic purple nonsulfur bacteria in a continuous co-culture system. Microbes Environ, 22, 82–87.

Papazi A., Makridis P., Divanach P., & Kotzabasis, K. (2008). Bioenergetic changes in the microalgal photosynthetic apparatus by extremely high CO2 concentrations induce an intense biomass production. Physiol. Plantarum, 132, 338-349.

Pawar, P.P., Odaneth, A.A., Vadgama, R.N., & Mallinath, A. (2019). Simultaneous lipid biosynthesis and recovery for oleaginous yeast Yarrowia lipolytica. Biotechnol Biofuels, 12, 237. https://doi.org/10.1186/s13068-019-1576-7.

Plaza, G.A., Chojniak, J., & Banat, I. M. (2014). Biosurfactant mediated biosynthesis of selected metallic nanoparticles. Int. J. Mol. Sci, 15, 13720–13737. doi: 10.3390/ijms150813720

Rane, A.N., Baikar, V.V., Kumar, V.R., & Deopurkar, R.L. (2017). Agro-Industrial Wastes for Production of Biosurfactant by Bacillus subtilis ANR 88 and Its Application in Synthesis of Silver and Gold Nanoparticles. Front. Microbiol, 8, 492. doi: 10.3389/fmicb.2017.00492

Romero, A.M., Ramos-Alonso, L., Montella-Manuel, S., Garcia-Martinez, J., de la Torre-Ruiz, M.A., Perez-Ortin, J.E., Martinez-Pastor, M.T., & Puig, S. (2019). A genome-Wide transcriptional study reveals that iron deficiency inhibits the yeast TORC1 pathway. Biochim. Biophys. Acta Gene Regul. Mech, 1862, 19441.

Rowe, G.E., Margaritis, A., & Dulmage, H.T. (1987). Bioprocess Developments in the Production of Bioinsecticides by Bacillus Thuringiensis. Crit Rev Biotechnol, 6(1), 87-127. https://doi.org/10.3109/07388558709086986.

Ruiz-Martinez, A., Garcia, N.M., Romero, I., Seco, A., & Ferrer, J. (2012). Microalgae cultivation in wastewater: nutrient removal from anaerobic membrane bioreactor effluent. Bioresour. Technol, 126, 247–253.

Santamaría, L., Reverón, I., de Felipe F.L., de Las Rivas, B., & Muñoz, R. (2018). Unravelling the Reduction Pathway as an Alternative Metabolic Route to Hydroxycinnamate Decarboxylation in *Lactobacillus plantarum*. Appl Environ Microbiol, 84(15), e01123-18. doi: 10.1128/AEM.01123-18.

Spagnuolo, M., Yaguchi, A., & Blenner, M. (2019). Oleaginous yeast for biofuel and oleochemical production. Curr Opin Biotechnol, 57, 73-81. doi: 10.1016/j.copbio.2019.02.011. DOI: 10.1016/j.copbio.2019.02.011

Sreeharsha, R.V., & Mohan, S.V. (2020). Obscure yet Promising Oleaginous Yeasts for Fuel and Chemical Production. Trends in Biotechnology, 38(8), 873-887. DOI:https://doi.org/10. 1016/j.tibtech.2020.02.004

Sullivan, D.P., Georgiev, A., & Menon, A.K. (2009). Tritium suicide selection identifies proteins involved in the uptake and intracellular transport of sterols in *Saccharomyces cerevisiae*. Eukaryot Cell , 8, 161–169.

Takahashi, J.A., Barbosa, B.V.R., Martins, B.d.A., Guirlanda, C.P., & Moura, M.A.F. (2020). Use of the Versatility of Fungal Metabolism to Meet Modern Demands for Healthy Aging, Functional Foods, and Sustainability. J. Fungi, 6, 223.

Tanaka, K., Natsume, A., Ishikawa, S., Takenaka, S., & Yoshida, K.I. (2017). A new-generation of Bacillus subtilis cell factory for further elevated scyllo-inositol production. Microb Cell Fact, 16, 67.

Tian, S., Ohta, A., Horiuchi, H., & Fukuda, R. (2018). Oxysterol-Binding protein homologs mediate sterol transport from the endoplasmic reticulum to mitochondria in yeast. J. Biol. Chem, 293, 5636–5648.

Venil, C.K., Velmurugan, P., Dufossé, L., Devi, P.R., & Ravi, A.V. (2020). Fungal Pigments: Potential Coloring Compounds for Wide Ranging Applications in Textile Dyeing. J. Fungi, 6, 68.

Vieira, É.D., Andrietta, M.d.G. S., & Andrietta, S.R. (2013).Yeast biomass production: a new approach in glucose-limited feeding strategy. Braz J Microbiol, 44(2): 551–558. doi:10.1590 /S1517-83822013000200035

Xavier, M.C.A., Coradini, A.L.V., Deckmann, A.C., & Franco, T.T. (2017). Lipid production from hemicellulose hydrolysate and acetic acid by Lipomyces starkeyi and the ability of yeast to metabolize inhibitors. Biochem Eng J, 118, 11–19.

Xiong, Y. M., D'Atri, J. J., Fu, S. L., Xia, H. P., & Seastedt, T.R. (2011). Rapid soil organic matter loss from forest dieback in a subalpine coniferous ecosystem. Soil Biol. Biochem, 43, 2450–2456. doi: 10.1016/j.soilbio.2011.08.013

Xu, K., Li, X-Q., Zhao D-L., & Zhang P. (2021). Antifungal Secondary Metabolites Produced by the Fungal Endophytes: Chemical Diversity and Potential Use in the Development of Biopesticides. Front. Microbiol, 12, 689527. | https://doi.org/10.3389/fmicb.2021.689527.

Yang, X.B., Jin. G.J, Gong, Z.W., Shen, H.W., Song, Y.H., Bai, F.W., & Zhao, Z.B.K. (2014). Simultaneous utilization of glucose and mannose from spent yeast cell mass for lipid production by Lipomyces starkeyi. Bioresour Technol, 158, 383–387.

Zhao, M., Guo, D.L., Liu, G.H., Fu, X., Gu, Y.C., Ding, L.S., & Zhou, Y. (2020). Antifungal halogenated cyclopentenones from the endophytic fungus *Saccharicola bicolor* of Bergenia purpurascens by the one strain-many compounds strategy. J. Agric. Food Chem, 68, 185–192. doi: 10.1021/acs.jafc.9b06594

Zhou, Q., Zhang, P., & Zhang, G. (2015). Biomass and Pigments Production in Photosynthetic Bacteria Wastewater Treatment: Effects of Light Sources, Bioresour Technol, 179, 505-509. DOI: 10.1016/j.biortech.2014.12.07.

Zhu, X., Zhou, D., Liang, F., Wu, Z., Shi, Z., & Li, C. (2017). Penochalasin K, a new unusual chaetoglobosin from the mangrove endophytic fungus *Penicillium chrysogenum* V11 and its effective semisynthesis. Fitoter, 123, 23–28. Doi: 10.1016/j.fitote.2017.09.016

Chapter-3

PRIMARY AND SECONDARY METABOLITES FROM BIOLOGICAL PROCESSES

Rajesh Kumar Srivastava[1], *Prakash Kumar Sarangi*[2*], *Akhilesh Kumar Singh*[3] *and Uttam Kumar Sahoo*[4]

[1]*Department of Biotechnology, GIT, Gandhi Institute of Technology and Management, Visakhapatnam, India*

[2]*College of Agriculture, Central Agricultural University, Imphal, Manipur, India.*

[3]*Department of Biotechnology, School of Life Sciences, Mahatma Gandhi Central University, Motihari, Bihar-845401, India.*

[4]*Department of Forestry, Mizoram University, Aizawl-796004, India*

**Corresponding author E mail: sarangi77@yahoo.co.in*

Abstract

Metabolites are organic compounds that are required for daily life of human. These compounds are normally synthesized by different types of microbial cells (bacteria, fungal, cyanobacterial or yeast cells), plant and animal cells that utilized the different metabolic pathways via consuming of different types of carbon and nitrogen sources. In ancient periods, different types of biological cells system were wild/natural nature strains that were less efficient and poor yields/titer in their bioprocesses. Now, due to advanced technology and scientific approaches those are available and also improved the metabolic pathways. These helped the microbial/others biological system more efficient that enhanced the productivity, titer and yield of different metabolites in biological /fermentation processes and helped in industrial operation/processes more beneficial in term of yields and profit. There are two types of metabolites produced via biological system; one is growth dependent metabolites (known as primary metabolites: fuels, organic acid or amino acids) and second one growth independent metabolites (known secondary metabolites: antibiotics/sterol/steroids). These metabolites played important to human life via helping in healthcare, pharmaceutical and medical sectors with human competent bioproducts. This chapter explores the different microbial systems, biological processes and its application in different sections/ industries.

Keywords: *Metabolites, Microbial system, Primary, Secondary, Metabolic pathways, Byproducts, Fermentation, Biological processes*

3.1. Introduction

Metabolites are normally bioproducts and also formed as intermediates in various cellular metabolisms. Metabolites showed multiples functions such energy conversion, signal and cofactor activity. Other function of metabolites is also noticed as epigenetic influences in various bioprocesses. Now, metabolites found application in clinical benefits/advantage with more valuable to health-care. Metabolites can be used as biomarkers in measurement of glucose in cerebrospinal fluids (CSF) and this is performed in cases of clinical suspicion of infection, inflammation or malignant processes (Wellen and Thompson, 2012).

Metabolites are biosynthesized as small molecules in biological cells and could be intermediate products in metabolic process with play important functions/role in biological system specially plant secondary metabolites. Primary metabolites are mostly found in many biological systems/ cell systems (Forgue et al., 2006). This metabolite can be known as central metabolites with its restricted presence to growing cells/ organisms. Ethanol, lactic acid, certain amino acid is a good example of primary metabolites, found mostly in microbial system that showed to biosynthesizing it (Forgue et al., 2006; Wellen and Thompson, 2012). Metabolites is formed and also concentrated in sees and vegetative storage organs. These metabolites are needed for physiological development especially their role in basic cell metabolic reactions. Normally, primary metabolites synthesized and also obtained from high plants for commercial application (Noda et al., 2017). But these metabolites are found in high volume and low value bulk chemicals. These plants originated metabolites can also be used as industrial raw materials, foods or food additives. Others plant products from plants are vegetable oils, fatty acids (uses for making of soaps and detergents) and carbohydrates (i.e., sucrose, starch, pectin and cellulose) (Wang et al., 2018). All these products, there are some products available that are more value and expensive primary nature of metabolites (such as myoinositol and β-carotene), But extraction, isolation and purification processes/ tasks of these products are difficult (Noda et al., 2017; Wang et al., 2018; Srivastava et al., 2020).

Other metabolites are found in secondary metabolites categories and these metabolites are diverse in structures and functions and it can function/serve as varied/ diverse weapons and can be applied against pathogenic bacteria, fungi, and plant or herbivorous animals with amoebae, and insects (Halouska, 2013). These secondary metabolites can be used in signaling pathways, also works as agents for symbiosis relationship between microbial system and plant, nematodes, insects and higher plant/ pollinators and also animals. These metabolites can be found as useful in fuels, medicines, cosmetics, food supplements and compounds. These metabolites have found wide range of applications in agriculture, ecology and healthcare section. Metabolomics is used to study the metabolic profiles (Markley et al., 2017). This branch of study is found to utilize the mass spectroscopy methods for analysis of different metabolites structure confirmation in biologic samples. And nuclear magnetic resonance is another approaches methods for estimation of metabolites information and used in metabolomics analysis (Halouska 2013; Markley et al., 2017).

Various types and versatile nature of research works were performed on metabolic pathways and also in determination of chemical structures of metabolites. Various book, review and research papers helped to explore/discuss the classification of alkaloids, based on ring skeleton and also its linkage of alkaloids clusters with synthetic routes/ synthesis pathways. Further discussion was various literatures were on structural similarity that based networks of metabolites and its uses/ application via establishing structure activity (Marshall et al., 2015). Final / end product/ metabolites with intermediates are synthesized/ formed during metabolic pathways/ process and is known as metabolites and primary metabolites are found to directly involve biological processes (Srivastava et al., 2020). These bioprocesses are growth, development and reproduction with capability to perform physiological functions in various biological systems under an intrinsic function (Marshall et al., 2015; Srivastava et al., 2020).

Secondary metabolites are reported as not essential metabolites as found for primary metabolites because it does not directly used/ involved in biological cell growth, development and reproduction tasks. But these are organic compounds, without directly involved in plant cells/ bodies survival tasks. But, these metabolites can produce some products that can id them in plant normal and development (Hasan et al., 2021). These compounds are biosynthetically derived from primary metabolites but it is more limited in distribution in the plant kingdom and also restricted to a particular taxonomic groups such as species, genus, family or closely related groups of family (Marshall et al., 2015; Hasan et al., 2021). This chapter explores the different types of primary metabolites, biosynthesis, pathways and their application in different (i.e., medical/ pharmaceutical) sectors/industries.

3.2. Metabolites and their Biosynthesis

Metabolites are bioproducts with organic compounds and are generally biosynthesized from various types of biological processes including fermentation (in batch, fed-batch and continuous modes), and biotransformation processes. And transketolase (*tkt*) enzymes deficient strain of *Bacillus pumilus* 21951(in ATCC known as *Bacillus subtilis*) is utilized in fermentation process that showed the capability to biosynthesize the D-ribose sugar as metabolites (can be primary category) and this sugar is found to some medical benefits and also utilized as precursor for some drugs preparation (such as anti-cancerous/anti-diabetics) (Bapat et al., 2006). This bacterial strain showed its capacity for D-ribose biosynthesis under an industrially viable process and this D-ribose is commercially important pentose sugars. Further, this *tkt* deficient *Bacillus* strain showed its direct effect on non-functional form/ stage in non-oxidative part of pentose phosphate pathways (PPP) and these changes in this PPP showed the unexpected physiological properties /characteristics of this *tkt* deficient *Bacillus* strain (Srivastava et al., 2009).

Next, effect of *tkt* enzymes deficient strain of *Bacillus pumilus* 21951 is found in decrease ability of utilization of D-glucose, altered carbon catabolites, lack of motility / spore formation. For *tkt* enzymes deficient strain of *Bacillus pumilus* 2195, morphological plasticity is reported and it showed its possible relation with D-productivity in fermentation process with a measurement of carbon flux via PPP of microbial system (Srivastava et al., 2009; Bapat et al., 2006). Normally this bacterial strain showed its dividing capability in nutrient rich media such as Luria-Bertani (LB) broth and its cells in elongated form (up to 20 times more) without any visible septum accompanied via moderate to high capability of D-ribose accumulation in glucose rich media/ production media. This bacterial cell (after taken from interval in fermentation) was stained with DAPI (4, 6-diamidino-2-phenylindole) and FtsZ antibody and it showed to confirm analysis the nucleoid separation and Z-ring formation in cells in LB media but not reported in cells of production media that contained high concentration of glucose substrate (Chauhan et al., 2006).

As we know that in normal cells of this bacteria that grown in LB, FtsZ protein is localized at the future division sites, and it forms a Z-ring at an early stage in cytokinesis of cell division. From this research work in 2009, this tkt bacterial strain is found to experience the inhibition or delay in Z-ring formation that causes the cell elongation in this cell. This bacterial also showed alter nature of cell membrane compositions due to tkt enzyme deficiency. And this published works claimed the hypothesis that due to PPP intermediates lacking, it showed two effects on the this microbial strain and these are first one: altered cell membrane led to delay in Z-ring formation and cell elongation and second one: is induction of genes of the oxidative parts of PPP in microbial strain (Chauhan et al., 2006; Bapat et al., 2006). And this has resulted in D-ribose accumulation in this bacterial strain. From this research works it can conclude that nutrient media like LB broth, found to alleviate these metabolites deficiencies thereby gaining/ restoring the normal cell division process in this strain and this has inhibited the excessive D-ribose accumulation. Further conclusion was on D-ribose production and cell elongation that not showed linear relationship with co-morbid nature. This reported works can help in designing in optimal media process and also monitoring strategy for fermentation that are based on morphological analysis task (Chauhan et al., 2006; Bapat et al., 2006; Srivastava et al., 2009).

In fermentation especially alcoholic fermentation process, yeast is found to effective microbial system with eukaryotic nature with its remarkable biodiversity and also extensive or more utilized in industrial operation/applications. In this regard, a very common species/ strain of yeast are *Saccharomyces cerevisiae* and this yeast strain is most widely usable microbial system in biotechnological process such as biochemical production/biosynthesis as successful operation/ application (Nandy and Srivastava, 2018). This yeast strain was used in biological conversion with the focus on the different utilities in renewable feedstock utilization into fuels energy and chemicals. This can be investigated in next step due to increased concern on sustainable issues at worldwide level. This yeast strain is used for application via comparing with its counterparts. Now, *Saccharomyces cerevisiae* is utilized as baker's yeast and is most usable microbial stain in industrial application with full details of genetic and physiological details/ information (Zhao et al., 2008).

Next this yeast is available in large collection of genetic tools. This yeast strain is found to more compatible of high density and large scale utility in fermentation processes. This yeast strain with optimal pathways was found suitable for variety of products. And *Saccharomyces cerevisiae* is found as one of most popular cell factories with successful usages in modern biotech industries (Hou et al., 2012). This yeast strain is found to biosynthesize the products such as ethanol, organic acid and amino acids. Next, utility of this stain is found to synthesize the enzyme and also therapeutic protein. This paper information was focused on different type of sustainable solutions, used to overcome the various environmental effects. This work information explored the targets of broad matrix of current advances and future prospect in yeast biotechnological information and application/ potential in different sectors (Zhao et al., 2008; Nandy and Srivastava, 2018) and figure 3.1 is showing the primary and secondary metabolites approaches.

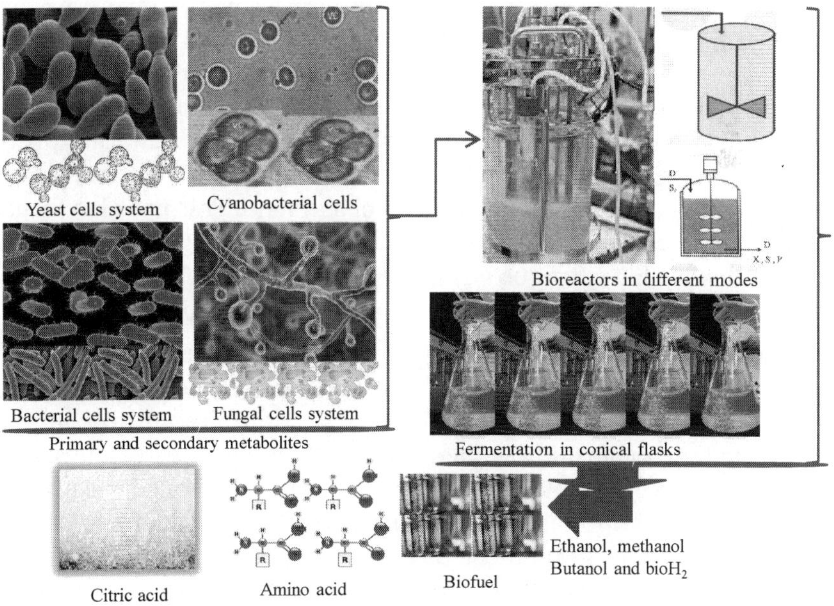

Figure 3.1- Primary and secondary metabolites synthesizing approaches from fermentation processes via using different microbial system.

3.2.1. Various Types of Metabolites

As it has been reported in many literatures that primary and secondary metabolites in plant, animal and many types of microbial system are founds with biosynthesis intra and extracellular nature from these biological system. And microbial cells system showed best production capabilities for many types of metabolites in their cells structures and these have explored at a commercial scales for getting the best yield and profits from these bio-products. As it has been noticed for high cost productions that are big challenges in most of the bioprocess industries. It needs to produce/ achieve at commercial scales with cheap production cost (Hou et al., 2012).

This issues is somewhat solved via developing and screening /using the engineered microbes that can be based on reconstruction model of genome-scale metabolite design/ theory. Now, many researchers worked on predicted genetic modification processes that can be applied for increased metabolites flux in biosynthetic pathways via helping in increased concentration/ quantity of desired end /final products (Wei et al., 2011).

3.2.2. Approaches of Microbial Strain Modification

In these efforts are shown in their genetic / pathways modification of microbial system and these caused the resultant microbial system for capability of converting a large quantity/ amount of carbon substrates into the expected desired products with minimal quantity of by-products biosynthesis in fermentation broth at optimal operating conditions (Zhang et al., 2019). For improvement of microbial system, metabolic engineering approach is found best tools via helping in improvement in metabolites yields, depending on the nature of the products, microbial cell design/ further modification tasks and also used substrates in fermentation processes. This published review paper has explored the comprehensive of state of art for various types of metabolites via microbial cells system. This has also focused on the selection approaches of efficient microbial strain and genetic components/ metabolic pathways modification/ engineering can be applied for effective one (Wei et al., 2011; Zhang et al., 2019).

Application of metabolic engineering is discussed for enable and also helped in development of microbial system/ strain with high performance and these microbes can be used for efficient production of natural and non-natural compounds/ chemicals from renewable nature of non-food biomasses or its residues. In recent period, with the help of the microbial system, various chemicals have produced in successful manners and it has experimentally proved with commercialized task for their production at large quantity. Still, researchers are trying their efforts for many types of chemical and materials that can be bio-based products with efficient production at cheap cost (Huccetogullari, 2019).

In this regard, numbers of aromatic chemical were typically derived from benzene, toluene and xylene components of petroleum industry and now they require using in large quantity/ amount in various industries. From last three decades, researchers have developed many metabolic pathways engineered microbial system that have applied for many types of biobased aromatic chemical biosynthesis and many types of these chemicals are derived from aromatic amino acid pathways (Du et al., 2018). Now many researchers have focused on highlight of latest metabolic engineering approaches and tools that are utilized for production of many types of aromatic chemicals. And many of these have derived from shikimate and aromatic amino acids such as L-phenylalanine, L-tyrosine, or L- tryptophan. And in coming period/ future, more and more numbers of engineered microbial systems will be developed with capabilities of efficient production of aromatic chemicals and it can further develop toward their industrial-scale biosynthesis from renewable biomass/ its waste matters (Du et al., 2018; Mora-Villalobos and Zeng, 2018).

3.3. Primary Metabolites

It is discussed in earlier sections in this report and other reports about primary metabolites nature. And primary metabolites are shown its nature for directly involving in processes of normal growth, development and reproduction of various types of biological systems (Kobayashi 2019). Primary metabolites shown its capability for performing a physiological functions in the biological cells under intrinsic functions and lactic acid and certain amino acids are the good examples of performing such tasks (Mutaguchi et al., 2013; Kobayashi 2019).

Other examples of primary metabolites are organic acid, nucleosides and also most of other amino acids and these products are simplest biotechnological products. Some primary metabolites can be used as raw materials for other processes such as fermentation and chemical syntheses (Taniguchi et al., 2017). Other properties of primary metabolites are shown via uses as nutritional supplement/favoring agents. These are rarely used as therapeutic compounds. Most of primary metabolites are produced at large scale/extensive level via microbial fermentation including yeast strains and these are mostly secreted into fermentation broth as extracellular products (Mutaguchi et al., 2018). These metabolites in fermentation processes are found in very concentration (that could be 1 to 100 $g.L^{-1}$) in most of biological agents in these processes (Taniguchi et al., 2017; Mutaguchi et al., 2018).

For recovery and purification processes of primary metabolites, there are relatively simple to complex purification steps with advanced developed flow sheets. In common process for these tasks in fermentation system via microbial/ yeast strains, solid-liquid separation can clarifies the supernatants via using a typical centrifugation processes and next step is to prepare for further process in recovery and purification (Vázquez et al., 2020). Next procedure is uses of ion-exchange chromatography technique, which capture the fermentative product. This technique uses the high-capacity resins. And direct crystallization of products can be done from conditional fermentative media (Gogami et al., 2011).

There are few concerns about products stability in process of different approaches in primary metabolites recovery and purification. It can decide the good product yields and process economics that are critical to design criteria. Primary metabolites utilization is dependent on purity criteria for food-grade application and it needed high level of purification that can be used in many industrial sectors (Aliashkevich et al., 2018). We discuss the higher level of purities of amino acids that uses in therapeutic applications especially as reagents for current GMP (cGMP) mammalian cell culture processes. And, United State Pharmacopeia, (USP), European Pharmacopeia (EP) and Japanese Pharmacopeia (JP) standard are needed to require the higher purities levels (up to 98.5%) (Cava et al., 2011; Aliashkevich et al., 2018)

In current period, primary metabolites (biosynthesized from microbial system) are used as essential nutrients and are more intensively studies with their engineering for respective microbial systems that can produce high yield, least amounts of byproducts and good productivity rate at cheap sources of feedstock that can reduce the product cost with easily providing of high profits to particular industries/ manufacturers (Wu et al., 2014). In this regard low mol. wt., metabolites are biosynthesized with important components of crop plants for both consumers and producers. And metabolic engineering approach can help to provide many essential nutrients such as amino acids, vitamins, lipid, and minerals in food sources (Waldie et al., 2014). Now researchers have worked on recent advances in the modification of a more specific secondary metabolism for the improvement in quality traits in any biological system (Wu et al., 2014; Hadacek and Bachmann, 2015).

And there are more highlights on essential amino acid biosynthesis and it is found to strictly regulate by some key enzymes such as dihydrodipicolinate synthase (DHDPS) enzyme in lysine (Lys) biosynthesis pathways. Next, enzyme is anthranilate synthase in tryptophan (Trp) biosynthetic pathways (Burrill et al., 2015). Nowadays folate has gained the considerable attention and interest among people and scientists that showed its value after it deficiency in biological system especially in human such as induction of neural tube defect during early pregnancy (Burrill et al., 2015; Lee et al., 2016). Next, metabolite is fatty acids and is another target for nutritional improvements and is critical in metabolic process, cardiovascular health, inflammatory responses and blood pressure regulation tasks. Next metabolite is chlorprothixene sulfoxide that comes from primary metabolite of chlorprothixene. This metabolite shows no neuroleptic action but it has the property of anticholinergic action (Undén et al., 1989).

Further, one case studies was discussed with two women doses effects that took the one doses (200 mg of chorprothixene by one women and second doses (200 to 400 mg of chorprothixene by another women) (Günther et al., 2018). And from these doses of chorprothixene concentration, reports has come on those women milk with those concentration of this compound their child's for this compound concentration was calculated to average concentrations of 2.4 μg/kg daily (with a maximum of 4.7 μg/kg daily) and the chlorprothixene sulfoxide doses (3.5 to 4.5 μg/kg daily) (Yu et al., 2020). This reports is discussed the fully breastfed instant and their infant can take in 0.2% of the maternal weight-related doses of active ingredient with babies studies with un- remarkable influence (Günther et al., 2018; Yu et al., 2020).

The products of amino acids are good examples with amenable to platform processing and it has reported in few production plants, used to produce multiple amino acids. In this regard, uses of various microbial systems such as *Corynebacterium, Bacillus, Aspergillium* or other species are used for biosynthetic pathways of primary metabolites and it was established from 1950 onwards for these metabolites (Dzugan et al., 2018). In ancient time/ periods, some technical problem as big challenges was found in case product recovery/ purification. Many types of unit operation were used in early processes to till today periods. And many places in the world, conduction of the academic research are now reported in regard to amino acid purification (Galili et al., 2016). Still, there is little insight into the proprietary research and this need to conduct within the major manufacturers of these amino acid products with many industrial application. Because it is technology base and needs to consider reaching a mature plateau (Galili et al., 2016; Dzugan et al., 2018).

Peptides and lectins are some groups of primary metabolites with exhibiting antimicrobial properties. These metabolites working mechanism of action is found due to formation of ion channels in cell membrane. Lectins examples such as ConA (isolated from *Canavalia ensiformis*) and ConBr (*Canavalia brasiliensis)* and these products shown its effects on retardation of biofilm development in pathogenic *Candida tropicalis* (Perçin et al., 2012; Turner and Butler, 2014). Next, metabolite is plant defensins (a cysteine-rich peptides) and it showed the antibiofilm activity against Candida species and it was further isolated as a novel antifungal peptides (Tn-AFP1 with mol. wt. 1230Da) from plant *Trapa natans* (Deng et al., 2005). This peptide exhibited a strong biofilm inhibitory activity against *Candida tropicalis* species. Similarly others metabolite is datusin (a peptide isolated from pericarp of *Datura strammnium* and this peptide showed fungicidal effects via weakening of the mature biofilm of *Candida ablican* (Giridhar and Srivastava, 2002; Deng et al., 2005).

Now a days, pomegranates (*Punica granatum* L) uses are reported as an important and interesting fruit tree and it is found to cultivate in many regions of the world. This plant cultivation and consumptions is found to increasing trends in many places due to dramatic increase in scientific interest in its biological science, methods of cultivation and adaption to environmental cues/adverse conditions. Next, this fruits is found to contain health-promoting properties (Akhavan et al., 2015). There are huge numbers of various types of metabolites that are produced in the pomegranate plants and these are determined with their contents in the bark, roots, leaves and fruits. In many reviews and research papers have discussed the polyphenolic with their compounds contents, antioxidant activity/ function and health-promoting compounds (Alcaraz-Mármola et al., 2017). And these compounds (such as organic acids, sugars, protein, lipids and amino acids) showed its reports with dedication to primary metabolites in pomegranate fruit. And special efforts were made to these recent studies with devoted to primary metabolites (Akhavan et al., 2015; Alcaraz-Mármola et al., 2017).

Many reports have shown on variation in the contents of primary metabolites within the pomegranate fruits and this was presented according to fruit tissues (peels, arils and sees), developmental stages of the fruits, environmental and climatic conditions and also genetics components background (Bar-Ya'akov et al., 2019). Most of reports on pomegranate fruits is based on metabolic contents, and contained no genetic/ molecular analysis except some work done on anthocyanins and hydrolysable tannins (pomegranate fruits are based on metabolic contents,) (Gavlighi et al., 2018). In places on this fruits studies, gene assignment and genetic control tasks were carried out in recent publication works and genome sequences from several pomegranate varieties and transcriptomics data from fruits, flower and leaves is discussed with expected facilitation for understanding of genetic control of metabolites in pomegranates (Gavlighi et al., 2018; Bar-Ya'akov et al., 2019).

Another metabolite was discussed as lovastatin and is known as anti-cholesterol medicine with many prescribed benefits for managing the cholesterol level in blood with minimum risks of suffering from heart-related disease. And very common fungal species is *Aspergillus terreus* (ATCC20542) and is shown its capability of production of lovastatin with supplementation of carbohydrates or sugar alcohols (Fang and Qu, 2018). Some research papers explored the works on application of metabolic engineering approach in fungal strain *A. terreus* that can re-route the precursor flow towards the lovastatin biosynthetic pathways that developed the capability of simultaneous mode of over-expression of gene for acetyl-CoA carboxylase (acc) to increase the precursor flux and climinate (+)-geodin biosynthesis (Milke and Marienhagen, 2020) .

This climinate (+)-geodin biosynthesis can compete the secondary metabolite with capability to remove the gene for emodin anthrone polyketide synthase (gedC). Further reports of metabolic engineering approach is found in alteration to metabolic flux in the double mutants (gedC D accox) strain and it was examined via checking the effects of using two different substrate formulations (Hasan et al., 2019). And this mutant strain (gedC D accox) was cultivated in a mixture of glycerol and lactose and showed significantly ($p<0.05$) increased the levels of metabolites precursors (such as malinyl-CoA~48% and acetyl-CoA~ 420%) that can be completely inhibited the (+)-geodin biosynthesis via increasing the level of lovastatin (152 mg/L that is 143% more titer) than the wild-type (WT) strain. Report supported the manipulation efforts of *A. terreus* strain metabolic biosynthetic pathway that can increase the efficiency of carbon flux towards lavastatin production (Hasan et al., 2018).

This effort can help in elevation of its overall production and also inabilities with uses of glycerol as a substrates source. This published works can provide a framework model for other medically or industrially important fungi to synthesize the valuable compounds sung sustainable carbon sources (Hasan et al., 2018; Hasan et al., 2019; Li et al., 2019)

3.4. Secondary Metabolites

Plants have capability to produce numerous secondary metabolites (SMs) during their growth stages and these secondary metabolites are found to serve various types of cellular functions that are needed and essential for physiological processes. In recent years, increased evidences have reported for some implicated stress and defense response signaling in their production phase (Isah, 2019). The types and concentration of secondary metabolites/molecules produced via a plant is dependent on their species, genotype, physiology developmental stages and also environmental factors during their growth periods. And physiological adaptation response by various plants taxonomic groups is found to coping with the stress and defense stimuli conditions (Walker et al., 2012).

And few recent years/ decades, there are numbers of witness that renewed the interest for studying the abiotics factors with influences to secondary metabolites production during the in vitro and also in vivo growth of plants (Plengmuankha et al., 2015). And currently number of molecular biology tools and techniques have been applied for facilitating the signaling processes and pathways that are involved in secondary metabolites production at sub-cellular, cellular, organ and also whole plants during in vivo and also in vitro growth conditions study via application of metabolic engineering approach in biosynthetic pathways of various intermediates (Isah, 2019; Plengmuankha et al., 2015).

As we know that secondary metabolites/compounds have shown any apparent function in plant's primary metabolism but these compounds/metabolites exhibited am ecological role such as involved in pollinators attractants and are representing in chemical adaptation to environmental stresses. Next, services are to perform chemical defenses against microbial system, insects, higher predators and also other plants influences (as allele-chemics). Secondary metabolites are normally accumulated by plants cells/complete plants in smaller quantities via compared to primary metabolites concentration/ concentration. And secondary metabolites synthesis in plants at larger quantity is very difficult to gain (Arrowsmith et al., 2012). So, secondary metabolites uses at commercials scale at larger concentration is found such as biologically active compounds and these are pharmaceuticals, flavor, fragrance and pesticides and these showed generally higher value-lower volume products via comparing to primary metabolites and these have trend to synthesize in specialized cell types or at distinct developmental phases/ stages (Kiselev et al., 2011).

These compounds can available via making their extraction and purification in easy and cost-effective ways. These metabolites are found to specialty/ fine chemicals once compared to primary metabolites (that synthesized in bulk quantity). Normally secondary metabolites are large size organic compounds and these need large numbers of specific enzymatic step for their production (Kiselev et al., 2011; Arrowsmith et al., 2012). In secondary metabolites groups, there are various steroids and alkaloid compounds especially some examples like steroidal sapogenins, digitalis glycosides, anticancer (*Catharanthus* or *Vinca*) alkaloid, belladonna alkaloids (i.e., atropine, hyocyamine and scopolamine) and cocaine (Jeong et al., 2013). Others secondary metabolites are reported such as colchicines, opium alkaloid (codenine, morphine and papaverine), physostigmine, pilocarpine with quinine, quinidine, reserpine and d- tubocurarine (Zeng et al., 2013).

Many other plants secondary metabolites are used in minimum/ limited quantities as pharmacological tools to study the various biochemical processes and its good examples are diterpene esters (among that are phorbol derivatives) and is come from lattices of various species of Euphorbia (it is members of the spurge family) and this found to potent irritants and co-carcinogene. This compound is found to use for study of chemical carcinogenesis (Gupta, 2013; Zeng et al., 2013).

As we are aware that different types of plants are found to continuous manner to expose to abiotics and biotics factors/stimuli and it can lead them to some wounding stress. To adapt such situation, different plant bodies have exhibited to develop the diverse defense mechanisms that help them to force/ signaling them for synthesizing the various secondary metabolites. And human can exploit these mechanisms to improve the efficacy of existing drugs and also to develop new one metabolite also (Dong et al., 2015). Many studies via literature and research works have focused on the effects of wounding stress on different plant parts such as leaves, stems and roots. And till current dates, there is no study that was investigated the accumulation of primary and galatamine contents, followed to exposure of a callus to wound stress. And some recent research works have been studied via exposing the *Lycoris radiat*e calli to wound stress and was assessed the expression levels of several genes (Wilson et al., 2012).

These genes were found to involve the metabolic pathways at various time points (from 0, 3, 6, 12..... to 72 and 96 hours) of exposures. Then these time points were quantified the primary and also galantamine contents via using the different analytical techniques such gas chromatography (GC) –time of flight mass spectroscopy (GC-TOFMS) and HPLC (high performance liquid chromatography) with also q-RT-PCR (quantitative real time-polymerase chain reaction) (Tian, et al., 2015). These analytical techniques helped in identification of eight genes of galantamine biosynthetic pathways such as LrPAL-2 and 3, LrC4H-2, LrC3H, LrTYDC2, LrN4OMT, LrNNR and LeCYP96T as a total gene). And out of total seven genes were significantly expressed via following the exposure to wounding stress except LrN4OMT gene. And galantamine contents of calli after completing of (from 3, 6, 12, 24, 48, 72 and 96hours) exposure, nearly 2.5, 2.5, 3.5, 3.5, 5.0 and 8.5 times higher than the 0 h of exposure respectively (Park et al., 2020). And further the totals of 48 hydrophilic metabolites were detected in the zero hour exposed callus and 96 h exposed callus via using the technique GC-TOFMS. Finally it was shown positive correlation between galantamine and initial precursor such as phenylalanine and tyrosine as detected (Tian, et al., 2015; Park et al., 2020) and some is shown in table 3.1.

Table 3.1 – Examples and benefits of primary metabolites.

Metabolites	Names	Microbial system/ benefits	References
Primary metabolites	Organic acids, sugars, protein, lipids and amino acids	Antioxidant activity/ function and health-promoting	Alcaraz-Mármola et al., 2017
Primary metabolites	Plant defensins (a cysteine-rich peptides	Antibiofilm activity against Candida species	Deng et al., 2005
Primary metabolites	Organic acid, nucleosides and also most of other amino acids	Simplest biotechnological products as nutritional supplementL / favoring agents	Taniguchi et al., 2017
Primary metabolites	Essential amino acid biosynthesis	anthranilate synthase in tryptophan (Trp)	Burrill et al., 2015
Primary metabolites	Lectins examples such as ConA	retardation of biofilm development in pathogenic Candida tropicalis	Perçin et al., 2012;
	Anthocyanins and hydrolysable tannins	Pomegranate fruits is based on these metabolic contents	Gavlighi et al., 2018
	Increased the level of lovastatin (152 mg/L that is 143% more titer) than the wild-type (WT) strain reports	It completely inhibited the (+)-geodin biosynthesis in A. terreus strain via metabolic engineered biosynthetic pathway	Hasan et al., 2018

3.5. Conclusion

This chapter explores about primary and secondary metabolites, biosynthesis, respective microbial system and also their applications in different sector. As it is mentioned that primary metabolites are growth dependent metabolites and produced by various types of microbial systems (its wild/ engineered strain) in different modes of fermentation. And secondary metabolites are non-growth dependent metabolite and is biosynthesized in late stationary phase of growth. Primary are produced in bulk quantity and comparatively less value added products than secondary metabolites. But secondary metabolites are high value –less volume bioproducts and are largely larger molecules. Nowadays, various approaches of metabolic pathway and genetic engineering approaches are applied to improve the microbial system efficiency and capability for better yield and productivity of primary and secondary metabolites. Primary metabolites are mostly biosynthesized by bacterial yeast and some fungal species with varying amounts depending on fermentation conditions. But fungal strain and some bacterial cells are found more efficient for secondary metabolites.

3.6. Abbreviations

Acc: acetyl-CoA carboxylase; **cGMP:** Current good manufacturing practices; **CSF:** Cerebrospinal fluids; **DHDPS:** Dihydrodipicolinate synthase; **EP:** European Pharmacopeia; **GC:** Gas chromatography; **GC-TOFMS :** GC –time of flight mass spectroscopy; **gedC:** emodin anthrone polyketide synthase; **HPLC:** High performance liquid chromatography; **JP:** Japanese Pharmacopeia; **LB:** Luria-Bertani; **Lys:** Lysine; **PPP:** Pentose phosphate pathways ; **q-RT-PCR:** Quantitative real time-polymerase chain reaction; **SMs:** Secondary metabolites; ***tkt*** **:** Transketolase; **Trp:** Typtophan; **USP:** United State Pharmacopeia; **WT:** Wild-type;

References

Akhavan, H., Barzegar, M., Weidlich, H., & Zimmermann, B.F. (2015). Phenolic compounds and antioxidant activity of juices from ten Iranian pomegranate cultivars depend on extraction. J. Chem, 907101. doi: 10.1155/2015/907101

Alcaraz-Mármola, F., Nuncio-Jáureguib, N., García-Sánchezc, F., Martínez-Nicolása, J.J. & Hernández, F. (2017). Characterization of twenty pomegranate (Punica granatum L.) cultivars grown in Spain: aptitudes for fresh consumption and processing. Sci. Hortic. 219, 152–160. doi: 10.1016/j.scienta.2017.03.008

Aliashkevich, A., Alvarez, L., & Cava, F. (2018). New insights into the mechanisms and biological roles of d-amino acids in complex eco-systems. Front. Microbiol, 9, 683. doi: 10.3389/fmicb.2018.00683

Arrowsmith, S., Egan, T.P., Meekins, J.F., Powers, D., & Metcalfe, M. (2012). Effects of salt stress on capsaicin content, growth, and fluorescence in a Jalapeño cultivar of Capsicum annuum (Solanaceae). Bios, 83(1), 1–8. https://doi.org/10.2478/s11756-019-00237-8.

Bapat, P.M., Das, D., Dave, N.N., & Wangikar, P.P. (2006). Phase shifts in the stoichiometry of rifamycin B fermentation and correlation with the trends in the parameters measured online. J Biotechnol, 127(1), 115–128

Bar-Ya'akov, I., Tian, L., Amir, R., & Holland, D. (2019). Primary Metabolites, Anthocyanins, and Hydrolyzable Tannins in the Pomegranate Fruit. Front. Plant Sci, 10, 620. doi: 10.3389/fpls.2019.00620

Burrill, J.S., Long, E.K., Reilly, B., Deng, Y., Armitage, I.M., Scherer, P.E., & Bernlohr, D.A. (2015). Inflammation and ER stress regulate branched-chain amino acid uptake and metabolism in adipocytes. Mol. Endocrinol, 29, 411–420.

Cava, F., Lam, H., de Pedro, M.A., & Waldor, M.K. (2011). Emerging knowledge of regulatory roles of d-amino acids in bacteria. Cell Mol. Life Sci, 68, 817–831. doi: 10.1007/s00018-010-0571-8

Chauhan, A., Madiraju, M., Fol, M., Lofton, H., Maloney, E., Reynolds, R., & Rajagopalan, M. (2006). Mycobacterium tuberculosis cells growing in macrophages are filamentous and deficient in FtsZ rings. J Bacteriol, 188(5),1856–1865.

Deng, M.-D., Severson, D.K., Grund, A.D., Wassink, S.L., Burlingame, R.P., & Berry, A. (2005). Metabolic engineering of Escherichia coli for industrial production of glucosamine and N-acetylglucosamine. Metab Eng, 7, 201–214.

Dong, Y.S., Duan, W.L., He, H.X.; Su, P., Zhang, M., Song, G.H., Fu, C.H., & Yu, L.J. (2015). Enhancing taxane biosynthesis in cell suspension culture of Taxus chinensis by overexpressing the neutral/alkaline invertase gene. Process Biochem, 50, 651–660.

Du, J., Yang, D., Luo, Z.W., & Lee, S.Y. (2018). Metabolic engineering of Escherichia coli for the production of indirubin from glucose. J. Biotechnol, 267, 19–28.

Dzugan, M., Wesołowska, M., Zaguła, G., & Puchalski, C. (2018). The comparison of the physicochemical parameters and antioxidant activity of homemade and commercial pomegranate juices. Acta Sci. Pol. Technol. Aliment, 17, 59–68. doi: 10.17306/J.AFS.0529

Fang, X., & Qu, Y. (2018). Fungal cellulolytic enzymes: Microbial production and application. Springer, pp. 1–282. https://doi.org/10.1007/978-981-13-0749-2

Forgue, P., Halouska, S., Werth, M., Xu, K., Harris, S., & Powers, R. (2006). NMR metabolic profiling of Aspergillus nidulans to monitor drug and protein activity. J Proteome Res, 5, 1916–1923.

Galili, G., Amir, A., & Fernie, A. (2016). The regulation of essential amino acid synthesis and accumulation in plants. Annu. Rev. Plant Biol, 67, 53–78. doi: 10.1146/annurev-arplant-043015-112213

Gavlighi, H.A., Tabarsa, M., You, S., Surayot, U., & Ghaderi-Ghahfarokhi, M. (2018). Extraction, characterization and immunomodulatory property of pectic polysaccharide from pomegranate peels: enzymatic vs conventional approach. Int. J. Biol. Macromol, 116, 698–706. doi: 10.1016/j.ijbiomac.2018.05.083

Giridhar, R.N., & Srivastava, A.K. (2002). Productivity improvement in l-sorbose biosynthesis by fed-batch cultivation of Gluconobacter oxydans. J Biosci Bioeng, 94, 34–38.

Gogami, Y., Okada, K., & Oikawa, T. (2011). High-performance liquid chromatography analysis of naturally occurring d-amino acids in sake. J. Chromatogr. B Analyt. Technol. Biomed. Life Sci, 879:3259–3267. doi: 10.1016/j.jchromb.2011.04.006.

Günther, K.N., Johansen, S.S., Wicktor, P., Banner, J, Linnet, K., 2018. Segmental Analysis of Chlorprothixene and Desmethylchlorprothixene in Postmortem Hair. J Analyt Toxicol. 42(9), 642-649. DOI:10.1093/jat/bky038.

Gupta P., 2013. Plant tissue culture of Stevia rebaudiana (Bertoni): a review. J Pharmacogn Phytother. 5(2), 26–33. https://doi.org/10.5897/JPP13.0258.

Hadacek, F., Bachmann, G., 2015. Low-molecular-weight metabolite systems chemistry. Front. Environ. Sci. 3, 12. doi: 10.3389/fenvs.2015.00012.

Halouska, S., Zhang, B., Gaupp, R., Lei, S., Snell, E., Fenton, R.J., Barletta, R.G., Somerville, G.A., Powers, R., 2013. Revisiting Protocols for the NMR Analysis of Bacterial Metabolomes. J Integr OMICS. 3, 120–137.

Hasan, H., Rahim M.H.A., Campbell, L., Carter, D., Abbas, A., & Montoya, A. (2018). Overexpression of acetyl-CoA carboxylase in Aspergillus terreus to increase lovastatin production. N Biotechnol, 44, 64-71. doi: 10.1016/j.nbt.2018.04.008.

Hasan, H., Rahim, M.H.A., Campbell, L, Carter, D., Abbas, A., & Montoya, A. (2021). Increasing Lovastatin Production by Re-routing the Precursors Flow of Aspergillus terreus via Metabolic Engineering, Molec Biotechnol. 10.1007/s12033-021-00393-w

Hasan, H., Rahim, M.H.A., Campbell, L., Carter, D., Abbas, A., & Montoy, A. (2019). Improved lovastatin production by inhibiting (+)- geodin biosynthesis in Aspergillus terreus. N Biotechnol, 52:19-24. doi: 10.1016/j.nbt.2019.04.003.

Hou, J., Tyo, K.E.J., Liu, Z., Petranovic, D., & Nielsen, J. (2012). Metabolic engineering of recombinant protein secretion by Saccharomyces cerevisiae. FEMS Yeast Res, 12, 490–510.

Huccetogullari, D., Luo, Z.W., & Lee, S.Y. (2019). Metabolic engineering of microorganisms for production of aromatic compounds. Microb. Cell Fact, 18, 41. https://doi.org/10.1186/s12934-019-1090-4

Isah, T. (2019). Stress and defense responses in plant secondary metabolites production. Biol Res, 52, 39. https://doi.org/10.1186/s40659-019-0246-3

Jeong, H., Lee, S.H., Yun, H.S., & Choi, S.R. (2013). Changes in allicin contents of garlic via light irradiation. Korean J Food Preserv, 20(1), 81–7. https://doi.org/10.11002/kjfp.2013.20.1.81.

Kiselev, K.V. (2011). Perspectives for production and application of resveratrol. Appl Microbiol Biotechnol, 90(2), 417–25. https://doi.org/10.1007/s00253-011-3184-8.

Kobayashi, J. (2019). d-Amino Acids and Lactic Acid Bacteria. Microorg, 7(12), 690. doi: 10.3390/microorganisms7120690.

Lee, C.C., Watkins, S.M., Lorenzo, C., Wagenknecht, L.E., Il'yasova, D., Chen, Y.I., Haffner, S.M., & Hanley, A.J. (2016). Branched-chain amino acids and insulin metabolism: The insulin resistance atherosclerosis study (IRAS). Diabetes Care, 39, 582–588.

Li, J., Zhang, Y., Li, J., Sun T., & Tian C. (2020). Metabolic engineering of the cellulolytic thermophilic fungus Myceliophthora thermophila to produce ethanol from cellobiose. Biotechnol Biofuels, 13, 23. https://doi.org/10.1186/s13068-020-1661-y

Markley, J.L., Brüschweiler, R., Edison, A.S., Eghbalnia, H.R., Powers, R., Raftery, D., & Wishart, D.S. (2017). The Future of NMR-Based Metabolomics. Curr Opin Biotechnol, 43, 34–40. doi: 10.1016/j.copbio.2016.08.001

Marshall, D.D., Lei, S., Worley, B., Huang, Y., Garcia-Garcia, A., Franco, R., Dodds, E.D., & Powers, R. (2015). Combining DI-ESI-MS and NMR datasets for metabolic profiling. Metabolomics, 11, 391–402.

Milke, L., & Marienhagen, J. (2020). Engineering intracellular malonyl-CoA availability in microbial hosts and its impact on polyketide and fatty acid synthesis. Appl. Microbiol. Biotechnol, 104(14), 6057–6065. https://doi.org/10.1007/s00253-020-10643-7

Mora-Villalobos, J.-A., & Zeng, A.-P. (2018). Synthetic pathways and processes for effective production of 5-hydroxytryptophan and serotonin from glucose in Escherichia coli. J Biol Eng, 12, 3.

Mutaguchi, Y., Kasuga, K., & Kojima, I. (2018). Production o d-Branched-Chain Amino Acids by Lactic Acid Bacteria Carrying Homologs to Isoleucine 2-Epimerase of Lactobacillus buchneri. Front Microbiol, 9, 1540. doi: 10.3389/fmicb.2018.01540.

Mutaguchi, Y., Ohmori, T., Akano, H., Doi, K., & Ohshima, T. (2013). Distribution of D-amino acids in vinegars and involvement of lactic acid bacteria in the production of D-amino acids. Springer plus, 691. doi: 10.1186/2193-1801-2-691. eCollection 2013.

Nandy, S.K., & Srivastava, R.K. (2018). A review on sustainable yeast biotechnological processes and applications. Microbiol. Res. 207, 83–90.

Noda, S., & Kondo, A. (2017). Recent advances in microbial production of aromatic chemicals and derivatives. Trends Biotechnol, 35, 785–96.

Park, C.H., Sathasivam, R., Nguyen, B.V., Baek, S.-A., Yeo, H.J., Park, Y.E., Kim, H.H., Kim, J.K., & Park, S.U. (2020). Metabolic Profiling of Primary Metabolites and Galantamine Biosynthesis in Wounded Lycoris radiata Callus, Plants, 9(11), 1616; https://doi.org/10.3390/ plants9111616.

Perçin, I., Yavuz, H., Aksöz, E., & Denizli, A. (2012). Mannose-specific lectin isolation from Canavalia ensiformis seeds by PHEMA-based cryogel. Biotechnol Prog, 28(3), 756-61. doi: 10.1002/btpr.1552.

Plengmuankhae, W., & Tantitadapitak, C. (2015). Low temperature and water dehydration increase the levels of asiaticoside and madecassoside in Centella asiatica (L.) Urban. South Afr J Bot, 97, 196–203. https://doi.org/10.1016/j.sajb.2015.01.013.

Srivastava, R.K., Akhtar, N., Vermam M., & Imandim S.B. (2020). Primary metabolites from overproducing microbial system using sustainable substrates. Biotechnol. Appl. Biochemis, 67(6), 852-874. https://doi.org/10.1002/bab.1927.

Srivastava, R.K., Jaiswal, R., Panda, D., & Wangikar, P.P. (2009). Megacell Phenotype and Its Relation to Metabolic Alterations in Transketolase Deficient Strain of Bacillus pumilus. Biotechnology and Bioengineering, 102(5), 1387-97. doi: 10.1002/bit.22184.

Taniguchi, M., Konya, Y., Nakano, Y., & Fukusaki, E. (2017). Investigation of storage time-dependent alterations of enantioselective amino acid profiles in kimchi using liquid chromatography-time of flight mass spectrometry. J Biosci Bioeng, 124(4), 414-418. doi: 10.1016/j.jbiosc.2017.04.019.

Tian, Y.Q., Zhang, C.Y., & Guo, M.Q. (2015). Comparative analysis of Amaryllidaceae alkaloids from three Lycoris species. Molec, 20, 21854–21869.

Turner S.A., & Butler G. (2014). The Candida Pathogenic Species Complex. Cold Spring Harb Perspect Med, 4(9), a019778. doi: 10.1101/cshperspect. a019778.

Undén, M., Nauntofte, B., & Dissing, S. (1989). Anticholinergic effects of cis-chlorprothixene characterized in rat parotid acini. Eur J Pharmakoi, 164(1), 129-38. doi: 10.1016/0014-2999(89)90239-2

Vázquez, J.A., Durán, A.I., Menduíña, A., Nogueira, M., Gomes, A.M., Antunes, J., Freitas, A.C., Dagá, E., Dagá, P., & Valcarcel, J. (2020). Bioconversion of Fish Discards through the Production of Lactic Acid Bacteria and Metabolites: Sustainable Application of Fish Peptones in Nutritive Fermentation Media. Foods, 9(9), 1239. doi: 10.3390/foods9091239.

Wal die, T., McCulloch, H., & Leyser, O. (2014). Strigolactones and the control of plant development: lessons from shoot branching. Plant J, 79, 607–622. doi: 10.1111/tpj.12488.

Walker, D.J., Martínez-Fernández, D., Correal, E., Romero-Espinar, P., & del Río, J.A. (2012). Accumulation of furanocoumarins by Bituminaria bituminosa in relation to plant development and environmental stress. Plant Physiol Biochem, 54, 133–9.

Wang, J., Shen, X., Rey, J., Yuan, Q., & Yan, Y. (2018). Recent advances in microbial production of aromatic natural products and their derivatives. Appl Microbiol Biotechnol, 102, 47–61.

Wei, S., Zhang, J., Liu, L., Ye, T., Gowda, G.A., Tayyari, F., & Raftery, D. (2011). Ratio analysis nuclear magnetic resonance spectroscopy for selective metabolite identification in complex samples. Anal Chem, 83, 7616–7623.

Wellen, K.E., & Thompson, C.B. (2012). A two-way street: reciprocal regulation of metabolism and signaling. Nat Rev Mol Cell Biol, 7, 13(4):270-6. doi: 10.1038/nrm3305.

Wilson, S.A., & Roberts, S.C. (2012). Recent advances towards development and commercialization of plant cell culture processes for the synthesis of biomolecules. Plant Biotechnol. J, 10, 249–268.

Wu, Q.-H., Wang, X., Yang, W., Nuessler, A. K., Xiong, L.-Y., Kuca, K., Dohnal, V., Zhang, X.-J., & Yuan Z.-H. (2014). Oxidative stress-mediated cytotoxicity and metabolism of T-2 toxin and deoxynivalenol in animals and humans: an update. Arch. Toxicol, 88, 1309–1326. doi: 10.1007/s00204-014-1280-0.

Yu, X., Li, X., You, S., Shi, Y., Zhu, R., Dong, Y., & Huang, C. (2020). Electro membrane extraction of chlorprothixene, haloperidol, and risperidone from whole blood and urine. J Chromatogr A, 1629, 461480. doi: 10.1016/j.chroma.2020.461480.

Zeng, J., Chen, A., Li, D., Yi, B., & Wu, W. (2013). Effects of salt stress on the growth, physiological responses, and glycoside contents of Stevia rebaudiana Bertoni. J Agric Food Chem, 61(24), 5720–6. https://doi.org/10.1021/jf401237x.

Zhang, S., Yang, W., Chen, H., Liu, B., Lin, B., & Tao, Y. (2019). Metabolic engineering for efficient supply of acetyl-CoA from different carbon sources in Escherichia coli. Microbial Cell Factories, 18(1), 1–11. https://doi.org/10.1186/s12934-019-1177-y.

Zhao, H.L., Xue, C., Wang, Y., Yao, X.Q., & Liu, Z.M. (2008). Increasing the cell viability and heterologous protein expression of Pichia pastoris mutant deficient in PMR1 gene by culture condition optimization. Appl. Microbiol. Biotechnol, 81, 235–241.

Chapter-4

FERMENTED FOOD AS SOURCE OF PROBIOTIC MICROORGANISMS

Th Anand Singh, Prakash K. Sarangi and Ng. Joykumar Singh

College of Agriculture, Central Agricultural University, Imphal, India-795004

****Corresponding author. Email: anan4thangjam@gmail.com***

Abstract

Probiotics, live microorganisms that confer health benefits to the host when consumed in adequate amounts, have gained significant attention due to their potential positive impact on human health. Fermented foods, a rich source of probiotics, have been consumed for centuries and are now recognized for their ability to improve gut health, enhance immune function, and promote overall well-being. This abstract aims to provide an overview of the role of fermented foods as a source of probiotics and their potential health benefits. Fermented foods are produced through the process of microbial fermentation, where beneficial microorganisms such as bacteria, yeasts, or molds transform the raw ingredients into a more nutritious and easily digestible form. These microorganisms contribute to the fermentation process by producing enzymes, organic acids, and various bioactive compounds, including probiotics. Probiotics present in fermented foods, such as Lactobacillus and Bifidobacterium strains, exert numerous beneficial effects on human health. Furthermore, the probiotic content of fermented foods can be influenced by various factors, such as fermentation duration, temperature, and starter cultures used. Optimizing these parameters can help enhance the probiotic load and maximize the health benefits of fermented foods. In conclusion, fermented foods serve as an abundant source of probiotics, offering numerous health benefits to individuals. Incorporating these foods into the daily diet can contribute to the maintenance of a healthy gut microbiota and overall well-being. Further research is warranted to explore the specific strains and mechanisms through which fermented foods exert their beneficial effects, allowing for targeted therapeutic interventions in various health conditions

Keywords: *Probiotics, microorganisms, fermentation, health, foods*

4.1. Introduction

Probiotics are groups of microorganisms which when consumed provide beneficial effects on the consumers (FAO/WHO 2001). Lactic acid-producing bacteria are the most common probiotics and are usually occurs naturally in yogurt, fermented milks or other fermented foods (Ahmed et al., 2013). Some of the beneficial effects of Probiotics are (i) improving intestinal tract health; (ii) enhancing the immune system, synthesizing and enhancing the bioavailability of nutrients; (iii) reducing symptoms of lactose intolerance, decreasing the prevalence of allergy in susceptible individuals; and (iv) reducing risk of certain cancers. Although the exact mechanisms of working of probiotics are largely unknown, but modifying gut pH, antagonizing pathogens through production of antimicrobial compounds, competing for pathogen binding and receptor sites as well as for available nutrients and growth factors, stimulating immunomodulatory cells, and producing lactase are suggested. In this last two decades or so important role of probiotics as a part of a healthy diet for human as well as for animals and may be an avenue to provide a safe, cost effective, and 'natural' approach that adds a barrier against microbial infection has been highlighted by many investigators around the world. Most important factors that affect a good probiotic candidate are its efficacy, selection criteria, food and supplement sources and its safety. Even the most effective probiotics in terms of dose, duration and efficiency necessitate strict scrutiny (Zommiti et al., 2020) to ensure safety and avoid probable side effects.

Fermented foods and beverages are known to improved shelf-life, safety, nutritional qualities, sensory, and functionality. It has also been established that the process of fermentation lead to formation of bioactive molecules, vitamins, and other attributes as well as augmented bio availability. Most fermented foods have live microorganisms in the category of GRAS (Generally regarded as safe) and having inherent properties of being a probiotic. The number of organisms in fermented foods varies considerably depending on the way the products were prepared and duration of storage. The commonly consumed fermented foods, including cultured dairy products, cheese, fermented sausage, fermented vegetables, soy-fermented foods, and fermented cereal products have been reported with potential source of probiotic microorganisms (Rezac et al., 2018). Many of these fermented foods contained 10^{5-7} lactic acid bacteria 10^{7-9} mesophillic bacteria and 10^{4-5} yeast per mL or gram, although there was considerable variation based on geographical region and sampling time. This chapter focus on the fermented foods and its natural microflora as source of Probiotics and related issues of probiotics as a health promoting supplements.

4.2. Types of Traditional Fermented Foods

There fermented foods are highly heterogenous in nature. Classification of the fermented foods is based on acidity of the foods as acidic, neutral or alkaline fermented foods or duration like long duration fermentation or short duration fermentation. But it is safer to classify based on its substrates. The major groups of substrates-based fermented foods are as follows:

- ☆ Fermented milk foods (*Dahi, Chhurpi*)
- ☆ Fermented cereal foods (*Dosa*)
- ☆ Fermented vegetable foods (*Soibum, Kimchi, sauerkraut*)
- ☆ Fermented soybean and seed foods (*Hawaijar*)
- ☆ Fermented fish products (*Ngari, Hentak*)
- ☆ Fermented beverages and Amylolytic starters (*Hamei, Atingba*)
- ☆ Miscellaneous fermented products (fermented meat, tea etc.) (Fig. 4.1)

Yogurt, kefir, and other cultured dairy product considered having good live microorganisms especially, *Streptococcus thermophilus* and *Lactobacillus delbrueckii subsp. Bulgaricus* though not restricted to this two and may harbour many other microorganisms. For many years, fermented dairy products were the most popular items with claims of live microorganisms until recently *Saurkraut* of Germany and *Kimchi* of Korea became highly sought after fermented food with active LAB.

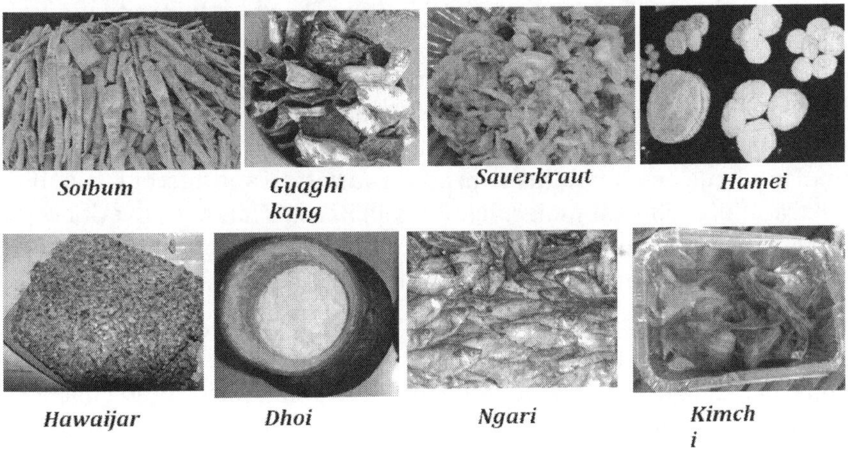

Figure 4.1 - Different types of fermented foods

4.3. Types of Microorganisms Associated with Fermented Foods and Probiotic

Many microorganisms in the fermented foods are unable to grow on conventional media and such microorganisms are known as unculturable microbes. Such microbes in fermented foods were identified indirectly using molecular. Whereas, those microorganisms isolated using standard known media with established techniques are called culturable microbes. Isolates were grouped and identified based on its phenotypic characters including its biochemical properties.

Further identification was subjected to genotypic methods where the genomes of the microbes were sequences and compared to establish the identity. Main types (with genera) of microorganisms associated with global fermented foods and beverages belong to Bacillus sp., Yeast and Lactic acid bacteria (LAB).

4.3.1. Bacillus

Bacillus signifies a Gram-positive, rod shaped, spore-forming, aerobic or facultative anaerobic bacterium. In general, the genus Bacillus is designated as a group of soil inhabitants. However, Bacillus spp. can be isolated from varied sources including air, water, human and animal gut, and also from vegetables and food (Alou et al., 2015; Singh et al., 2014). Nonetheless, Bacillus spp. have been the most heterogeneous group both phenotypic and genotypic characteristics. Some distinct species have also been recognized as opportunistic pathogen or toxin producer in human or animal hosts. The genus Bacillus is closely related to Lactobacillus spp., the distinguished candidate probiotic. Both belong to same class, Bacilli under the phylum Firmicutes.

The enhanced tolerance and survivability in harsh environment of gastrointestinal tract and its inherent ability of bacillus in production of good numbers of secretory proteins, enzymes, vitamins, antimicrobial compounds and carotenoids made them a favourite workhorse in functional food research. Besides, these spore formers are more stable throughout processing and storage of food and pharmaceutical preparations. Further, Bacillus strains also possess biotherapeutic potential by virtue of their ability to produce variety of antimicrobial peptides and other extracellular effector molecules. But still Bacillus has been overlooked due to its association of some diseases. Hence, proper identification and characterisation of beneficial Bacillus spp. and substantiation of GRAS status is required (Zommiti et al., 2020).

Bacillus has not gained high popularity in research and commercialisations like its counterpart LAB despite having comparable probiotic attributes. Functionalities of several Bacillus spp. have been validated in several *in vitro* and *in vivo* models (Elshaghabee et al., 2017). Studies have shown that Bacillus spp. offers higher acid tolerance and more stability during heat processing and low temperature storage (Bader et al., 2012). Besides they possess food fermentation abilities (Terlabie et al., 2006) along with anti-oxidant, antimicrobial, immuno-modulatory propertiesd (Ripert et al., 2016). Few Bacillus spp. particularly, *B. cereus, B. weihenstephanensis, B. anthracis*, and *B. thuringiensis* species are well known to produce various toxins, including enterotoxin and exotoxins. Hence, the beneficial spore former probiotic strains have safety concerns and therefore, the GRAS grade Bacillus that have been consumed in the form of fermented foods since long time (Tamang et al., 2016) must be the first choice for probiotics. Table 4.1 depicts different types of fermented foods of the world with its associated microorganisms

Table 4.1 - Different types of fermented foods of the world with its associated microorganisms

Sl. No.	*Fermented food*	*Substrate*	*Nature*	*Microorganisms*	*Country*	*Reference*
	Chhurpi	Yak/cow milk	Cheese-like product, soup, curry, pickle	*Lb. farciminis, Lb. casei, Lb. biofermentans, W. confusus*	India, Nepal, Bhutan, China (Tibet)	Tamang, 2010
	Dahi	Cow/buffalo milk, starter culture	Curd, savory	*Lb. bifermentans, Lb. alimentarius, Lb. paracasei, Lact. Lactis, Strep. Cremoris, Strep. Lactis, Strep. Thermophilus, Lb.bulgaricus, Lb. acidophilus, Lb. helveticus, Lb. cremoris, Ped. Pentosaceous, P. acidilactici, W. cibara, W. paramesenteroides, Lb. fermentum, Lb. delbrueckii subsp. Indicus, Saccharomycopsis sp., Candida sp.*	India, Nepal, Sri Lanka, Bangladesh, Pakistan	Harun-ur-Rashid et al., 2007; Patil et al., 2010
	Ham	Cured pork	Semi-dry, breakfast	LAB, yeasts, micrococci	Spain, Italy	Simoncini et al., 2007
	Hamei	Rice	Starter to make atingba	*Mucor sp., Rhizopus sp., Sacch. Cerevisiae, Pic. Anomala*	India	Jeyaram et al., 2009
	Hawaijar	Soybean	Alkaline, sticky	*B. subtilis, B. licheniformis, B. amyloliquefaciens, B. cereus, Staph. aureus, Staph. sciuri, Alkaligenes* sp., *Providencia rettgers, Proteus mirabilis*	India	Jeyaram et al., 2008; Singh et al., 2014
	Jeot kal	Fish	High-salt fermented, staple	LAB	Korea	Guan et al., 2011
	Kefir	Goat, sheep, cow milk	Alcoholic fermented milk, effervescent milk	*Tor. holmii, Tor. delbruechii, Lb. brevis, Lb. caucasicus, Strep. thermophilus, Lb. bulgaricus, Lb. plantarum, Lb. casei, Lb. brevis*	Russia, Europe	Bernardeau et al., 2006

Sl. No.	Fermented food	Substrate	Nature	Microorganisms	Country	Reference
	Kimchi (beachoo)	Cabbage, green onion, hot pepper, ginger	Acidic, mild-sour, wet, side-dish	*Leuc. mesenteroides, Leuc. citreum, Leuc. gasicomitatum, Leuc. kimchii, Leuc. inhae, W. koreensis, W. cibaria, Lb. plantarum, Lb. sakei, Lb. delbrueckii, Lb. buchneri, Lb. brevis, Lb. fermentum, Ped. acidilactici, Ped. pentosaceus*	Korea	Nam et al., 2009
	Ngari	Fish (*Puntius sophore*),	Fermented fish	*Lact. Lactis, Lb. plantarum, Ent. Faecium, Lb. fructosus, Lb. amylophilus, Lb. coryniformis, B. subtilis, B. pumilus, Micrococcus sp., Candida sp., Saccharomycopsis sp*	India	Devi et al., 2015
	Sauerkraut	Cabbage	Acidic, sour, wet, salad, side dish	*Leuc. mesenteroides, Ped. pentosaceus; Lb. brevis, Lb. plantarum, Lb. sakei*	Europe, USA, Canada, Australia	Johanningsmeir et al., 2007
	Soibum	Bamboo shoot	Acidic, sour, soft, curry	*Lb. plantarum, Lb. brevis, Lb. coryniformis, Lb. delbrueckii, Leuc. Fallax, Leuc. Lact. Lactis, Leuc. Mesenteroides, Ent. Durans, Strep. Lactis, B. subtilis, B. lichniformis, B. coagulans, B. cereus, B. pumilus Pseudomonas fluorescens, Saccharomyces sp., Torulopsis sp*	India	Jeyaram et al., 2010 Tamang et al., 2012
	Togwa	Cassava, maize, sorghum, millet	Fermented gruel or beverage	*Lb. brevis, Lb. cellobiosus, Lb. fermentum, Lb. plantarum and Ped. pentosaceus, Candida pelliculosa, C. tropicalis, Issatchenkia orientalis, Sacch. cerevisiae*	Tanzania	Mugula et al., 2003

4.3.2. Yeast

Few yeast have been claimed to have probiotic potentials. *Sachharomyces boulardii* is the most important and most studied probiotic yeast (Pais et al., 2020). Several studies conducted with *S. boulardii* for the treatment and prevention of various forms of diarrhoea has promising results. Encouraging research perspectives have been opened in terms of maintenance, treatment of inflammatory bowel diseases. The clinical activity of *S. boulardii* has been documented in antibiotic-associated diarrhoea and recurrent *Clostridium difficile* intestinal infections (Pais et al., 2020). The mechanism of action, although not entirely conclusive, of *S. boulardii* was partially elucidated (Czerucka et al., 2007). Many researches has demonstrated that *S. boulardii* have specific probiotic properties and herald the new therapeutic uses of this yeast as 'probiotic.

4.3.3. Lactic Acid Bacteria (LAB)

LAB are group of microaerophillic, non spore forming Fermicutes. They are gram-positive, lack cytochromes and porphyrins and are thus, catalase and oxidase negative. Some may take up oxygen through the mediation of flavo-protein oxidase and/or re-oxidize NADH produced during the carbohydrate metabolisms (Adams and Moss, 1999). They are usually associated with many of the fermented foods of the world and are the most important microorganisms as the source of probiotics due to its innate GRAS grade properties and other desirable properties.

LAB Probiotics have been easily accepted by consumers as a health promoting than other microorganisms. LAB as probiotics for both human and animal use has been reported (Yoon et al., 2004; Savadogo et al., 2006). *Lactobacillus* and *Bifidobacteria* species were themost common probiotic for human consumption. Particularly, *Bifidobacterium animalis, Bifidobacterium lactis, Bifidobacterium bifidum, Bifidobacterium breve* and *Bifidobacterium longum* biotypes *infantis* and *longum* were frequently implemented in probiotic products (Masco et al., 2005). *Lactobacillus casei* is the most widely studied among lactobacilli for probiotic applications. It has been studied particularly for its effects on various types of diarrhea and its modulating capacity of the immune system. *Bifidobacteria* are another species LAB probiotics that is commonly isolated from feaces of humans, animals, birds and are present in high numbers in breastfed babies (Mitsuoka, 1992). *Bifidobacteria* has many metabolic properties such as being fermentative and producing lactate, among other acids (Vankerckhoven et al., 2008). *Lactobacilli* on the other hand, are rods, long and slender and form chains in many cases (Frazier and Westhoff, 2002).

4.4. Health Benefits of Probiotics

Many studies have revealed the health benefits of probiotics to human as well as animals. Due to its easier administration and relatively cheaper option than many antibiotics and other therapies, it is becoming popular in recent times. Health benefits of probiotics could be local (in gastrointestinal tract; GIT) or systemic (whole body).

4.4.1. Localised Effect

GIT diseases that have been shown to respond to probiotics are infectious diarrhoea, gastric ulcers caused by Helicobacter pylori, irritable bowel syndrome, antibiotic-associated diarrhoea, Clostridium difficile-associated disease, inflammatory bowel disease, traveler's diarrhoea, necrotizing enterocolitis, constipation, lactose intolerance, and colorectal cancer (CAST 2007). The most common pathogen are adenovirus, enterovirus, enterotoxigenic, *E. coli*, *Salmonella, Shigella, Yersinia, Campylobacter, Vibrio cholera, Cryptosporidium*, and *Giardia* are also common (Allen et al., 2004) of which rotaviruses are the most rampant. Probiotics reduced the acute diarrhoea on the third day; the duration of diarrhoea was also shortened by 30 to 48 hours (Allen et al., 2004). *L. reuteri* DSM 17938 could improve the acute diarrhoea in hospitalized children and decreased the duration and frequency of diarrhoea (Francavilla et al., 2012).

It may be worth mentioning here that other than the pathogens affecting diarrhoea in patients, antibiotic treatment may also cause diarrhea by upsetting the microbial balance. Probiotics (*L. rhamnosus* and *Saccharomyces boulardii*) prevent antibiotic associated diarrhea in children by restoring the gut microbiota (Johnston et al., 2011). A systematic evaluation of the efficacy of probiotics such as *Lactobacillus, Bifidobacterium, Saccharomyces, Streptococcus, Enterococcus*, and *Bacillus* for the prevention or treatment of diarrhoea indicated reduced risk of antibiotic-associated diarrhoea (Hempel et al., 2012). *S. boulardii* was effective against traveler's diarrhoea, nutrition-related diarrhoea, and *H. pylori* related symptoms and promising results were also observed in *C. difficile* associated disease, irritable bowel syndrome, acute adult diarrhoea, Crohn's disease, giardiasis, and HIV-related diarrhoea (McFarland et al 2010). Ulcerative colitis and Crohn's disease are inflammatory bowel diseases (IBD) manifested by disproportionate host-microbes interactions which could later develop into colorectal cancer (Geier et al 2007). Although no specific pathogen has been determined as a causative agent, probiotics could be effective to prevent or treat IBD by balancing the gut microbiota.

A concoction of probiotics consisting of *L. paracasei, L. plantarum, L. acidophilus, L. delbrueckii subsp. bulgaricus, B. longum, B. brevis, B. infantis*, and *Streptococcus thermophilus* was effective in ulcerative colitis in children (Miele et al 2009). Probiotics holds promise for the treatment of IBD despite limited clinical evidence. Therapeutic treatment of Irritable bowel syndrome (IBS) is not available currently and the best approach for an effective treatment might be restoration of the microbiota composition, which is an additional factor affecting IBS pathogenesis (Lee et al., 2011). Probiotics have been shown to decrease pathogenic translocation after injury. *L. plantarum* L2 when administered in adequate quantity prevented the bacterial infection in mesenteric lymph node, spleen, liver, and kidney by reducing the secretion of pro-inflammatory cytokines and endoepithelial cell apoptosis (Wang et al., 2011).

4.4.2. Systemic Effects

Probiotics reduce the incidence of cold and flu, ear-nose-throat infection, urogenital infection, atopic dermatitis, allergic rhinitis, or used as a vaccine adjuvant (Leyer et al.,2009, Zommiti et al., 2020). Studies have shown that rhinitis, rhinosinusitis, rhinopharyngitis or common cold, pharyngitis, epiglottitis and laryngitis, were relieved after probiotic administration. This approach is effective alternative to antibiotic treatment which could lead to resistance in pathogens (Popova et al., 2012). Common viral pathogens that cause respiratory tract infection are rhinoviruses, coronaviruses, parainfluenza and influenza while *Streptococci, Mycoplasma pneumoniae, Chlamidia pneumoniae, Corynebacterium diphteriae, Staphylococcus aureus*, and *Streptococcus pneumonia* are bacterial causative agents. *L. rhamnosus* and *L. acidophilus* were considered most effective against such pathogens while other species such as *L. delbrueckii, L. paracasei, L. plantarum, L. casei, L. helveticus* and *B. bifidum* were other probiotics shown to be effective (Popova et al., 2012). The low cost and easy administration of probiotics have been considered one of the major advantages over antibiotic treatment for ventilator-associated pneumonia in an intensive care unit. Administration of many LAB probiotics resulted in reduce incidence of this upper respiratory tract infection and decreased the length of intensive care unit stay; also lessen or avoid *Pseudomonas aeroginosa* colonization (Siempos et al., 2010). Uropathogenic bacteria such as *E. coli, Klebsiella, Proteus*, and *Enterobacter* cause urinary tract infection, and antibiotic treatment may also effect kidney. An alternative approach would be administration of probiotics to counter the growth of the pathogen. A mouse model of urinary tract infection using *L. plantarum* LPLM-O1 as a prophylactic therapy decrease *E. coli* without any adverse effects (de Arellano et al., 2012). Recent studies suggested role of host immune status and GIT microbiota in probiotic efficacy. Probiotics were also shown to be effective in treatment of allergies (Guvenc et al., 2016). Numerous studies envisage the potential of probiotics as vaccine adjuvants by boosting the serum antibodies and improving immunization (Rezac et al., 2018). Few studies reported the induction of neutralizing antibodies by administering probiotic such as *L. rhamnosus* GG or *L. acidophilus* CRL431 (de Vrese et al., 2005) in case polio virus. The increase in circulating antibodies would neutralize the virus and thus avert the disease (Kemgang et al., 2014). Table 4.2 depicts health benefits of probiotics

Table 4.2 - Health benefits of probiotics

Sl No.	*Disorder/health benefits*	*Probiotic microorganism*	*Reference*
	Obesity and weight related issues	*Lactobacillus acidophilus* La_5, *Bifidobacterium animalis* Bb_{12}	Ivey et al., 2015.
		Weissella koreensis	Le et al., 2018
	Type 2 diabetes	*Lactobacillus acidophilus* La5, *Bifidobacterium animalis* Bb_{12}	Ejtahedet al., 2012
	Dyslipidemia	*Lactobacillus acidophilus, Lactobacillus reuteri* NCIMB	Shimizu et al., 2015

Sl No.	Disorder/health benefits	Probiotic microorganism	Reference
	Constipation	*Bifidobacterium lactis*	Dimidi et al., 2014
	Enterocolitis	*Lactobacillus* *Bifidobacterium*	AlFaleh et al., 2014
	Antibiotic associated Diarrhoea	*Bacillus clausii* UBBC-07	Sudha et al., 2019
	Lactose intolerance	*Bifidobacterium animalis* DSM 26137, *Lactobacillus plantarum* DSM 26329	Roškar et al., 2017
	Allergic Rhinitis	*Streptococcus paracasei*-33	Peng, et al., 2015
	Blood pressure	*Lactobacillus helveticus* and *Saccharomyces cerevisiae*	Hendijani et al., 2018
	Atopic Dermatitis	*Lactobacillus rhamnosus* HN001	Wickens et al., 2013.
	Cancer	*Lactobacillus paracasei*	Tarrah et al., 2019
	Mental disorder	*Lactobacillus, Bifidobacterium*	Mayer et al., 2015

4.5. Mechanism of Action

The available literatures suggested the mechanism of probiotic action are either probiotic or host can interact directly (e.g. host-microbe interaction) or via intermediates (e.g. microbe-microbe interaction) to bring forth the health benefits. Probiotics generate beneficial effects by producing antimicrobials, competing with pathogens for nutrients and adhesion sites, inducing host defence secretion, modulating immune responses, enhancing intestinal barrier function, or altering the gut microbiota (Ng et al., 2009). Mechanisms of action of probiotics guide further research on probiotics to select probiotic strains for explicit purposes like prophylactic and therapeutic therapies.

Almost all clinical trials used were generally recognized as safe (GRAS) category microorganisms (Zommiti et al., 2020), but opportunistic infections may result from probiotic administration to immune-compromised or critically ill people. A mixture of *L. acidophilus, L. casei, L. salivarius, Lactococcus lactis, B. bifidum,* and *B. lactis* increased enterocyte damage and bacterial translocation in patients with acute (Besselink et al 2009). Hence, probiotics are not recommended for patients with acute pancreatitis (Witteman et al., 2008). Similarly, *S. boulardii*, a nonpathogenic yeast, was responsible for about 3.6% cases of fungemia in France (Piarroux et al., 1999). Rarely, *L. rhamnosus, L. casei*, and *L. paracasei* were also isolated from blood of patients after organ transplantation (0.241%). *B. brevis* was also found to be a causative agent of neonatal meningitis (Nakazawa et al., 1996) which normally showed no side effects but gained more weight and had fewer abnormal abdominal signs when administered (Kitajima et al., 1997). In addition, probiotics when given to patients in intensive care unit have fewer incidence of ventilator-associated pneumonia, lesser period of stay in ICU, and infection by *Pseudomomas aeroginosa* in the respiratory tract (Siempos et al., 2010).

4.6. Conclusion

Probiotics are live microorganisms which, when consumed in sufficient amounts, bestow health benefit to the host. But the dead cells and parts of cells have also been shown to have positive effects, primarily immunomodulatory activity. Fermented foods have been a reservoir of diverse micro flora. Lactic acid bacteria, yeast and Bacillus have been the dominant microbes in many fermented foods. Their inherit properties made them ideal for exploiting as probiotics. Health benefits are specific to microbial strains and are dose dependent. While everyone believed probiotics have health benefits, their mechanisms of action are yet to answer satisfactorily. The in vitro systems commonly employed to study probiotic mechanisms of action were however frequently failed to predict in vivo efficacy. Thus, the food particularly traditional fermented foods origin GRAS grade probiotics hold healthier implication than those developed from other sources.

Template

References

Adams, M.R., & Moss, M.O. (1999). The Royal Society of Chemistry, Thomas Graham House, Science Park, Cambridge CB4 0WF. UK. Food Microbiology.

AlFaleh, K., & Anabrees, J. (2014). Probiotics for prevention of necrotizing enterocolitis in preterm infants. Evid. Based Child Health, 9, CD005496.

Allen, S.J., B. Okoko, E., Martinez, G., Gregorio, & Dans, L. F. (2004). Probiotics for treating infectious diarrhoea. Cochrane Database Syst Rev, CD003048.

Alou, M.T., Rathored, J., Khelaifia, S., Michelle, C., Brah, S., & Diallo, B. A., et al. (2015). Bacillus rubiinfantis sp. nov. strain mt2T, a new bacterial species isolated from the human gut. New Microbes New Infect, 8, 51–60. doi: 10.1016/j.nmni.2015.09.008.

Bader, J., Albin, A., & Stahl, U. (2012). Spore-forming bacteria and their utilization as probiotics. Benef. Microbes, 3, 67–75. doi: 10.3920/BM2011.0039.

Bernardeau, M., Guguen, M., & Vernoux, J.P. (2006). Beneficial lactobacilli in food and feed: Long-term use, biodiversity and proposals for specific and realistic safety assessments. FEMS Microbiology Review 30: 487–513.

Besselink, M.G., van Santvoort, H.C., Renooij, W., M.B., de Smet, M.A., Boermeester, K. & Fischer, et al. (2009). Intestinal barrier dysfunction in a randomized trial of a specific probiotic composition in acute pancreatitis. Ann Surg, 250:712-719.

CAST. 2007. Probiotics: their potential to impact human health, vol. 36. Council for Agricultural Science and Technology (CAST), Ames, IA.

Czerucka, D., Piche, T., Rampal, P. (2007). Review article: yeast as probiotics – Saccharomyces boulardii Aliment Pharmacol Ther, 26, 767–778. doi:10.1111/j.1365-2036.2007.03442. x.

de Arellano, A.R., Sanchez, M., Vera, R., Jara, S., Gonzalez, M., & Castro, E. (2012). Effect of orally-administered Lactobacillus plantarum LPLM-01 strain in an immunosuppressed mouse model of urinary tract infection. Benef Microbes, 3:51-59.

de Vrese, M., Rautenberg, P., Laue, C., Koopmans, M., Herremans, T., & Schrezenmeir, J. (2005). Probiotic bacteria stimulate virus-specific neutralizing antibodies following a booster polio vaccination. Eur J Nutr, 44:406-413.

Devi K.R., Deka, M., Jeyaram, K. (2015) Bacterial dynamics during yearlong spontaneous fermentation for production of ngari, a dry fermented fish product of Northeast India Intl. J Food Microbiol, 199:62-71.DOI: 10.1016/j.ijfoodmicro.2015.01.004.

Dimidi, E., Christodoulides, S., Fragkos, K.C., Scott, S.M., & Whelan, K. (2014). The effect of probiotics on functional constipation in adults: A systematic review and meta-analysis of randomized controlled trials. Am. J. Clin. Nutr, 100, 1075–1084.

Ejtahed, H.S., Mohtadi-Nia, J., Homayouni-Rad, A., Niafar, M., Asghari-Jafarabadi, M., & Mofid, V. (2012). Probiotic yogurt improves antioxidant status in type 2 diabetic patients. Nutrition, 28, 539–543.

Elshaghabee, F.M.F., Rokana, N., Gulhane, R.D., Sharma, C., & Panwar, H. (2017). Bacillus As Potential Probiotics: Status, Concerns, and Future Perspectives. Front. Microbiol, 8:1490. doi: 10.3389/fmicb.2017.01490

FAO/WHO. (2001). Health and nutritional properties of probiotics in food including powder milk with live lactic acid bacteria. Report of a joint FAO/WHO expert consultation.

Francavilla, R., Lionetti, E., Castellaneta, S., Ciruzzi, F., Indrio, F., & Masciale, A., et al. (2012). Randomised clinical trial: Lactobacillus reuteri DSM 17938 vs. placebo in children with acute diarrhoea - a double-blind study. Aliment Pharmacol Ther, 36:363-369.

Geier, M. S., Butler, R.N., & Howarth, G.S. (2007). Inflammatory bowel disease: current insights into pathogenesis and new therapeutic options; probiotics, prebiotics and symbiotic. Int J Food Microbiol, 115:1-11.

Guan, L., Cho, K.H., & Lee, J.H. (2011). Analysis of the cultivable bacterial community in jeotgal, a Korean salted and fermented seafood, and identification of its dominant bacteria. Food Microbiology, 28: 101–113.

Guvenc, I.A., Muluk, N.B., Mutlu, F.S., Eski, E., Altintoprak, N., Oktemer, T., & Cingi, C. (2016). Do probiotics have a role in the treatment of allergic rhinitis? : A comprehensive systematic review and meta-analysis. Am. J. Rhinol. Allergy, 30, 157–175.

Harun-ur-Rashid, M., Togo, K., Ueda, M., & Miyamoto, T. (2007). Probiotic characteristics of lactic acid bacteria isolated from traditional fermented milk "Dahi" in Bangladesh. Pakistan Journal of Nutrition, 6: 647–652.

Hempel, S., Newberry, S.J., Maher, A.R., Wang, Z., Miles, J.N., & Shanman, R., et al. (2012). Probiotics for the prevention and treatment of antibiotic-associated diarrhoea: a systematic review and meta-analysis. Jama, 307:1959-1969.

Hendijani, F., & Akbari, V. (2018) Probiotic supplementation for the management of cardiovascular risk factors in adults with type II diabetes: A systematic review and meta-analysis. Clin. Nutr, 37, 532–541.

Ivey, K.L., Hodgson, J.M., Kerr, D.A., Thompson, P.L., Stojceski, B., & Prince, R.L. (2015). The effect of yogurt and its probiotics on blood pressure and serum lipid profile; a randomized controlled trial. Nutr. Metab. Cardiovasc. Dis, 25, 46–51.

Jeyaram, K., Singh, W.M., Premarani, T., Devi, A.R., Selina Chanu, K., Talukdar, N.C., & M. Rohinikumar Singh. (2008). Molecular identification of dominant microflora associated with "Hawaijar" – a traditional fermented soybean (Glycine max L.) food of Manipur, India. International Journal of Food Microbiology 122: 259–268

Jeyaram, K., Romi, W., Singh, T.A., Devi, A.R., & Devi, S.S. (2010). Bacterial species associated with traditional starter cultures used for fermented bamboo shoot production in Manipur state of India. International Journal of Food Microbiology, 143: 1–8.

Jeyaram, K., Singh, W.M., Capece, A., & Romano, P. (2008). Molecular identification of yeast species associated with "Hamei"—A traditional starter used for rice wine production in Manipur, India. International Journal of Food Microbiology, 124: 115–125.

Johanningsmeier, S., McFeeters, R.F., Fleming, H.P., & Thompson, R.L. (2007). Effects of Leuconostoc mesenteroides starter culture on fermentation of cabbage with reduced salt concentrations. Journal of Food Science, 72: M166–M172.

Kemgang, T.S., Kapila, S., Shanmugam, V.P., & Kapila, R. (2014). Cross-talk between probiotic lactobacilli and host immune system. J. Appl. Microbiol, 117, 303–319.

Kitajima, H., Sumida, Y., Tanaka, R., Yuki, N., Takayama, H., & Fujimura, M. (1997). Early administration of Bifidobacterium breve to preterm infants: randomised controlled trial. Arch Dis Child Fetal Neonatal Ed, 76:F101-F107.

Le, B., & Yang, S.H. (2018). Isolation of Weissella strains as potent probiotics to improve antioxidant activity of salted squid by fermentation. J. Appl. Biol. Chem, 61, 93–100.

Lee, B.J., & Bak, Y.T. (2011). Irritable bowel syndrome, gut microbiota and probiotics. J Neurogastroenterol Motil, 17:252-266.

Leyer, G. J., Li, S., Mubasher, M.E., Reifer, C., & Ouwehand, A.C. (2009). Probiotic effects on cold and influenza-like symptom incidence and duration in children. Pediatrics. 124:e172-e179. 37.

Masco, L., Huys, G., De Brandt, E., Yemmerman, R., & Swings, J. (2005) Culture-dependent and culture-independent qualitative analysis of probiotic products claimed to contain Bifidobacterial. Int. J. Food Microbiol, 102: 221-230.

Mayer, E.A., Tillisch, K., & Gupta, A. (2015). Gut/brain axis and the microbiota. J. Clin. Investig, 125, 926–938.

McFarland, L.V. (2010). Systematic review and meta-analysis of Saccharomyces boulardii in adult patients. World J Gastroenterol, 16:2202-2222.

McFarland, L.V., & Dublin, S. (2008). Meta-analysis of probiotics for the treatment of irritable bowel syndrome. World J Gastroenterol, 14:2650-2661.

Miele, E., Pascarella, F., Giannetti, E., Quaglietta, L., Baldassano, R.N. & Staiano, A. (2009). Effect of a probiotic preparation (VSL#3) on induction and maintenance of remission in children with ulcerative colitis. Am J Gastroenterol, 104:437-443.

Mitsuoka, T. (1992). The Human Gastrointestinal Tract, In: Wood BJB (Ed). The Lactic acid bacteria. vol 1 Elservier science. Pub. London.

Frazier, W., & Westhoff, D. (2002). Food Microbiology. Tata McGrawHill Publishing Company Limited, New Delhi.

Mugula, J.K., Ninko, S.A.M., Narvhus, J.A., & Sorhaug, T. (2003). Microbiological and fermentation characteristics of togwa, a Tanzanian fermented food. International Journal of Food Microbiology, 80: 187–199.

Nakazawa, T., Kaneko, K., Takahashi, H., & Inoue, S. (1996). Neonatal meningitis caused by Bifidobacterium breve. Brain Dev, 18:160-162.

Nam, Y-D., Chang, H.W., Kim, K.H., Roh, S.W., & Bae, J.W. (2009). Metatranscriptome analysis of lactic acid bacteria during kimchi fermentation with genome-probing microarrays. International Journal of Food Microbiology, 130: 140–146.

Ng, S.C., Hart, A.L., Kamm, M.A., Stagg, A.J. & Knight, S.C. (2009). Mechanisms of action of probiotics: recent advances. Inflamm Bowel Dis, 15:300-310.

Patil, M.M., Pal, A., Anand, T., & Ramana, K.V. (2010). Isolation and characterization of lactic acid bacteria from curd and cucumber. Indian Journal of Biotechnology, 9: 166–172.

Peng, Y., Li, A., Yu, L., & Qin, G. (2015). The role of probiotics in prevention and treatment for patients with allergic rhinitis: A systematic review. Am. J. Rhinol. Allergy, 29, 292–298.

Piarroux, R., Millon, L., Bardonnet, K., Vagner, O., & Koenig, H. (1999). Are live saccharomyces yeasts harmful to patients? Lancet, 353:1851-1852.

Popova, M., Molimard, P., Courau, S., Crociani, J., Dufour, C., Le Vacon, F., & Carton, T. (2012). Beneficial effects of probiotics in upper respiratory tract infections and their mechanical actions to antagonize pathogens. J Appl Microbiol. doi: 10.1111/j.1365-2672.2012.05394. x.

Rezac, S., Kok, C.R., Heermann, M., & Hutkins, R. (2018). Fermented Foods as a Dietary Source of Live Organisms. Front. Microbiol, 9:1785. doi: 10.3389/fmicb.2018.01785.

Ripert, G., Racedo, S. M., Elie, A.M., Jacquot, C., Bressollier, P., & Urdaci, M.C. (2016). Secreted compounds of the probiotic Bacillus clausii strain O/C inhibit the cytotoxic effects induced by Clostridium difficile and Bacillus cereus toxins. Antimicrob. Agents Chemother, 60, 3445–3454. doi: 10.1128/AAC.02815-15.

Roškar, I., Švigelj, K., Štempelj, M., Volfand, J., Štabuc, B., Malovrh, Š., & Rogelj, I. (2017). Effects of a probiotic product containing Bifidobacterium animalis subsp. animalis IM386 and Lactobacillus plantarum MP2026 in lactose intolerant individuals: Randomized, placebo-controlled clinical trial. J. Funct. Foods, 35, 1–8.

Salminen, S., von Wright, A., Morelli, L., Marteau, P., Brassart, D., de Vos, W.M., Fonden, R., Saxelin, M., Collins, K., Mogensen, G., Birkeland, S.E. & Mattila-Sandholm, T. (1998). Demonstration of safety of probiotics -- a review. Int J Food Microbiol, 44:93-106.

Savadogo, A., Quattara, C.A., Bassole, H.N., Traore, S.A. (2006). Bacteriocins and Lactic acid bacteria a mini review. Afr. J. Biotechnol, 5(9): 678-684.

Shimizu, M., Hashiguchi, M., Shiga, T., Tamura, H.O., Mochizuki, M. (2015). Meta-Analysis: Effects of Probiotic Supplementation on Lipid Profiles in Normal to Mildly Hypercholesterolemic Individuals. PLoS ONE, 10, e0139795.

Siempos, I.I., Ntaidou, T.K. & Falagas, M.E. (2010). Impact of the administration of probiotics on the incidence of ventilator-associated pneumonia: a meta-analysis of randomized controlled trials. Crit Care Med, 38:954-962.

Simoncini, N., Rotelli, D., Virgili, R., & Quuintavalla, S. (2007). Dynamics and characterization of yeasts during ripening of typical Italian dry-cured ham. Food Microbiology, 24(6): 577–584.

Singh, T.A., Devi, K.R., Ahmed, G., & Jeyaram, K. (2014). Microbial and endogenous origin of fibrinolytic activity in traditional fermented foods of Northeast India. Food Research International, 55: 356–362.

Sudha, M.R., Jayanthi, N., Pandey, D.C., & Verma, A.K. (2019). Bacillus clausii UBBC-07 reduces severity of diarrhoea in children under 5 years of age: A double blind placebo-controlled study. Benef. Microb, 10, 1–6.

Tamang, J.P., Watanabe, K., & Holzapfel, W. H. (2016). Review: diversity of microorganisms in global fermented foods and beverages. Front. Microbiol, 7:377. doi: 10.3389/fmicb.2016.00377.

Tamang, J.P. (2010). Himalayan Fermented Foods: Microbiology, Nutrition, and Ethnic Values. New: CRC Press, Taylor & Francis Group.

Tamang, J.P. (2012). Chapter 4. Plant-based fermented foods and beverages of Asia. In: Handbook of Plant-Based Fermented Food and Beverage Technology, Second Edition (Eds: Hui, Y.H. and Özgül, E.). CRC Press, Taylor & Francis Group, New York, pp. 49–90.

Tarrah, A., Da Silva Duarte, V., de Castilhos, J., Pakroo, S., Junior, W., Luchese, R., Guerra, A.F., Rossi, R.C., Ziegler, D.R., & Corich, V., et al. (2019). Probiotic potential and biofilm inhibitory activity of Lactobacillus casei group strains isolated from infant feces. J. Funct. Foods, 54, 489–497.

Terlabie, N. N., Sakyi-Dawson, E., & Amoa-Awua, W. K. (2006). The comparative ability of four isolates of Bacillus subtilis to ferment soybeans into dawadawa. Int. J. Food Microbiol, 106, 145–152. doi: 10.1016/j.ijfoodmicro. 2005.05.021.

Vankerckhoven, V., Huys, G., Vancanneyt, M., Vael, C., Klare, I., Romond, M., Entenza, J., Moreillon, P., Wind, R., Knol, J., Wiertz, E., Pot, B., Vaughan, E., Kahlmeter, G., & Goossens, H. (2008). Biosafety assessment of probiotics used for human consumption: recommendations from the EUPROSAFE project. Trends Food Sci. Technol, 19: 102-114.

Wickens, K., Stanley, T.V., Mitchell, E.A., Barthow, C., Fitzharris, P., Purdie, G., Siebers, R., Black, P.N., & Crane, J. (2013). Early supplementation with Lactobacillus rhamnosus HN001 reduces eczema prevalence to 6 years: Does it also reduce atopic sensitization? Clin. Exp. Allergy, 43, 1048–1057.

Yoon, Y.D., Kyung, Y.Y., & Woodams, E.E. (2004). Probiotication of Tomato Juice by Lactic Acid Bacteria. J. Microbiol. 42(4): 315-318.

Zommiti, M., Feuilloley, M.G.J., Connil, N. (2020). Update of Probiotics in Human World: A Nonstop Source of Benefactions till the End of Time, Microorganisms, 8, 1907. doi:10.3390/microorganisms 8121907.

Chapter-5

FERMENTED ETHNIC FOODS AND BEVERAGES: A STRATEGY OF FOOD PRESERVATION AND ETHNOMEDICINAL IMPORTANCE

Akanksha Tiwari[1], Soumitra Tiwari[1], Saurabh Saxena[2], Vinit Singh Baghel[3] and Dhananjay Shukla[3]*

[1] Atal Bihari Vajpayee University, Bilaspur Chhattisgarh, India

[2] Nextgen Invitro Diagnostics, Gurugram, Haryana, India

[3] Department of Biotechnology, Guru Ghasidas University, Bilaspur, Chhattisgarh, India

**Corresponding Author Email: soumitratiwari@bilaspuruniversity.ac.in*

Abstract

Practically in every civilization on every continent, fermented foods and alcoholic drinks have long been a staple of the human diet. These foods are frequently well-preserved and provide reliable supplies of proteins, vitamins, minerals, and other nutrients. Despite these similarities, there are numerous changes in substrates and products, as well as the types of microorganisms used in the production of fermented foods and drinks across the world. Lactic acid bacteria, Aspergillus spp., Penicillium spp., and Mucorales moulds, and yeasts, particularly Saccharomyces spp., are the most commonly employed microorganisms. Fermented foods are prevalent in global diets with dairy, beverage and cereal goods dominating. Fermented foods and drinks make up a large component of all diets across the world, accounting for around one-third of total food intake and making a considerable nutritional, flavour and interest contribution to human food consumption. Fermented foods have different health benefits like lowering the blood cholesterol level, increased immunity, anti-inflammatory actions, protection against bacteria, etc.

Keywords: *Fermentation, Lactic acid bacteria, Microbes, Consumption, Preservatives.*

5.1. Introduction

Several scholars claim that the term "fermentation" derives from the Latin word "fermentare," which means "to leaven," while others say it comes from the Latin verb "fervere," which means "to boil." Several meanings have been used to describe not only the principle of fermentation but also its implementations, based on these Latin words so far. As a result, the word "fermentation" was often used in the early days of the term. Fermentation, according to some more recent meanings, is a mechanism in which a substance is broken down into a simpler substance, typically as a result of the metabolic activity of microorganisms such as yeast or bacteria. Fermentation processes, in the sense of physiology and biochemistry, are typically metabolic pathways that result in changes in organic molecules due to the action of microorganisms and/or enzymes (Fruton, 2006).

Microorganisms have long been thought to be the only organisms capable of carrying out fermentation, yet fermentation is prevalent in almost all the living world including bacteria, archaea and eukaryotes. Fermentation is a biological process that involves extracting energy from molecules. As a result of the degradation of organic nutrients, fermentation produces ATP (usually under anoxic conditions). Fermentation takes place in the gastrointestinal tracts of animals, mainly, due to microbial flora (Fruton, 2006; Savage, 1986).

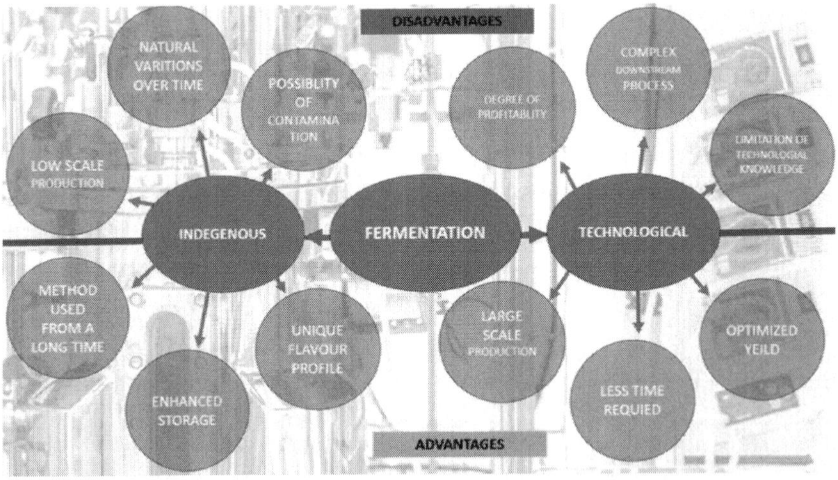

Figure 5.1. - Fermentation types and their advantages and disadvantages.

Over time, the goals of adopting fermentation-based technologies have changed. As a result, fermentation has been widely used in all cultures around the world to enhance food preservation for shorter periods or to create beverages or regional dishes that include fermented sauces, vegetables, meat, or fish. The term "indigenous fermentation" or "classical fermentation" refers to this form of fermentation (it is used to produce foods and beverages using natural microbes following traditional procedures based on cultural practices). Many products have been standardised and commercialised by indigenous fermentation (ales, natural yeast; cheeses, natural fungi; wines, natural yeast), while many others are produced in small amounts for specialised makers or remain uncommercialised (products of indigenous/local cultures). This is the case for kefir, kimchi, sauerkraut and other fermented foods (Hesseltine et al., 1980; Zhu et al., 2013). Figure 5.1. depicts the fermentation types and their advantages and disadvantages.

Humans have been fascinated by fermentation-based processes since antiquity because of their possible applications. The sources of fermentation are unknown; however, it is thought that the first fermentation mechanism was inadvertently discovered when salt was added to food. Fermentation has been traced back to 7000 BC by a few experts. Fermented foods such as beer, wine, leavened bread (formed primarily by yeasts) and cheeses have been produced in Western locations since Neolithic times (made by bacteria and moulds). These communities along with techniques were quickly adopted by East Asian locations (6000–1500 BC), where local processes were followed to manufacture fermented foods, yoghurt and other fermented milk products, pickles, sauerkraut, vinegar (soured wine), butter, and a range of typically alcoholic beverages (McGovern et al., 2004; Vouillamoz et al., 2006; Cavalieri et al., 2003).

Humans have been consuming fermented foods for thousands of years. While they were most likely created for preservation, it would have been obvious that these foods had other desirable qualities. Fermented foods have distinct tastes, textures, appearances and functions when compared to the raw materials from which they are produced. Fermented foods may have been deliberately developed many years ago, long before nutrition science, as a reliable source of vitamins, minerals, calories and other nutrients (Steinkraus, 1994).

Fermented foods and beverages are an important element of Indian culture. Fermentation is one of the most ancient and cost-effective ways of food production and preservation (Jeyaram, 2009). It has been used by ancient men as a cost-effective way of extending the life of fermented goods since ancient times (foods and beverages). Traditional fermentation of foods is one of the world's oldest biotechnological methods (Sekar and Mariappan, 2007). Fermented foods can be produced naturally or with the addition of a starting culture containing effective microorganisms that transform the substrates into edible foods that are ethnically and socially acceptable to the locals.

Primitive people used a variety of storage methods to store excess food from any source, particularly foods that are temporary and have a short shelf life (perishable). People eat about 5,000 distinct fermented foods around the world today, many of which are indigenous and manufactured in small quantities to fulfil the needs of certain communities (Ray et al., 2016).

Other fermented foods, on the other hand, tend to have evolved and originated in specific places. Milk from cows, horses, goats and other livestock, for example, was readily available everywhere as pastoral farming practices and animal husbandry were prevalent. Cultured milk, cheese and fermented dairy products arose in the Middle East, Europe and India as a result. Animal agriculture, on the other hand, was restricted in China, Japan, Korea and other Far East countries. The types of substrates used to manufacture fermented foods and alcoholic beverages were determined by social, cultural, religious and economic factors.

Reduced blood cholesterol levels, increased immunity, protection against bacteria, battling carcinogenesis, osteoporosis, diabetes, obesity, autism, atherosclerosis and alleviating lactose sensitivity symptoms are all benefits of fermented foods (Tamang and Kailasapathy, 2010). The health advantages associated with fermented meals are also attributed to bioactive peptides produced during the microbial breakdown of proteins by bacteria involved in fermentation (Hebert et al., 2010; Martinez Villaluenga et.al., 2017; Otag and Hayta, 2013; Walther and Sieber, 2011). The most well-known subtype of bioactive peptides is angiotensin-1-converting enzyme (ACE) inhibitor peptides, which are formed when milk proteins are destroyed by proteinases during fermentation present in lactic acid bacteria's cell walls. Because of the known antihypertensive effects of these peptides, particularly Valyl-prolyl-proline (VPP) and isoleucyl-prolyl-proline (IPP), fermented dairy products are advised as a non-pharmacological approach for the treatment of hypertension (Beltran-Barrientos et al., 2016; Nejati et al., 2013; Usinger et.al., 2009).

The natural fermentation mechanism, which involves pathogenic, non-functional and functional microorganisms, produces the vast majority of fermented items (Tamang et al., 2016; Tamang et al., 2015). Lipase, glucoamylase, protease and amylase are all enzymes made by functional microorganisms. When microorganisms are combined with enzymes, they boost up the nutrient content of anti-nutritive and inedible substrates converting them into nutritious products that can provide a range of health benefits to the consumers (Tamang and Samuel, 2010; Tamang et al., 2016). The shift from hunting and gathering to organised food processing is thought to have arisen in the Middle East around 10000-15000 years ago. About 5000 years ago, records of barley being fermented into beer and grapes being fermented into wine have been found (Borgstrom, 1968).

Fermentation has been mostly used to improve food's shelf life while also enriching its flavour and texture (Hutkins, 2018). Fermented foods with living microbes have lately become a popular dietary approach for improving human health (Marco et al., 2017). The most frequent bacteria found in fermented foods are lactic acid bacteria (LAB) from various genera, such as *Lactobacillus, Streptococcus* and *Leuconostoc*, although other bacteria, yeast, and fungus also play a role. Probiotic bacteria are also carried by commercially processed fermented foods. It remains unclear which fermented foods contain living bacteria and how these microbes alter the gut microbiome (Slashinski et al., 2012).

While yeasts are required for fermentation in bread, beer, wine and distilled alcoholic beverages, the production organisms are either inactivated by heat (in the case of some beers) or physically removed by filtration or other means (in the case of wine and beer). Furthermore, many fermented foods undergo heat treatment after fermentation to improve food safety or prolong shelf life. As a result, fermented sausages are often cooked after fermentation and thermal processing is used to make many fermented vegetables shelf-stable.

Several fermented foods like pickles, for example, are placed in brines containing salt and organic acids rather than being fermented. In addition, because of unfriendly climatic conditions that limit microbial populations over time, fermented foods that have not been thermally processed may contain low quantities of live or viable organisms. The absence of live microbes in the final product, however, does not rule out a positive functional role (Bhatia and Paliwal, 2010). Microbes in food fermentation, for example, may generate vitamins or other beneficial compounds in the presence or inhibit anti-nutrients, but these substances may not be available at the time of intake (Rezac, et al., 2018). Figure 5.2. depicts an overview of the fermented food market and estimated growth by year 2027. The fermented food market was valued at 565.09 billion USD in 2019 and is forecasted to reach 875.21 billion USD in 2027 with the compound annual growth rate (CAGR) of 6.0%. The key factors leading to the accelerated expansion of the market are demand of higher shelf-life food, urbanization and awareness among people about the fermented foods (Data obtained from https://www.emergenresearch.com)

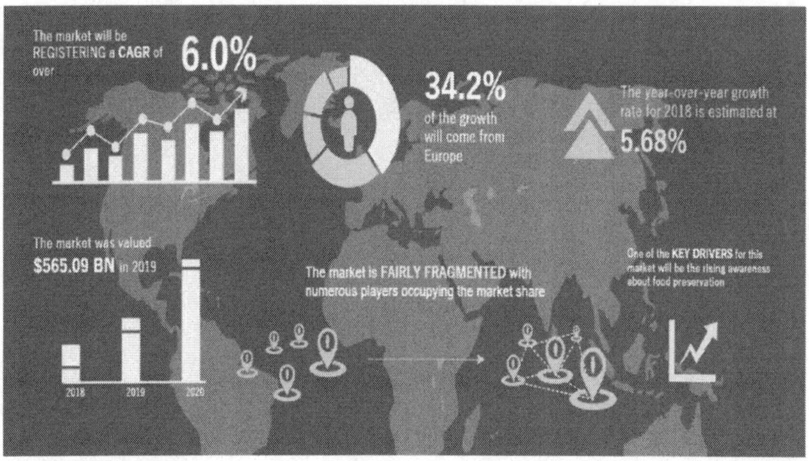

Figure 5.2. - An overview of the fermented food market and estimated growth by year 2027 (Courtesy - https://www.emergenresearch.com).

5.2. Types of Fermented Foods/Beverages

5.2.1. Fermented foods/beverages made from grains

Grains are India's primary source of nourishment. The consumption of grains in India's various cultures can be seen in the literature. Due to the presence of starch and proteins, grains have a high calorific value. Some of the grains have antinutritional qualities, such as the presence of phytate and trypsin inhibitors. Antinutritional factors may be removed using processing methods such as fermentation, germination, soaking, and cooking (Egounlety and Aworh, 2003).

Although grains lack nutrients due to the lack of essential amino acids and other basic compounds, fermented grains are a simple way to improve nutritional value sensory and functional qualities (Ozdemir et al., 2007). Fermentation reduces the content of carbohydrates, indigestible polysaccharides and oligosaccharides in grains while increasing the synthesis of certain amino acids and the bioavailability of vitamin B (Blandino et al., 2003).

Fermentation also decreases the number of non-nutrients like polyphenols, phytates and tannins while increasing the number of nutrients like free amino acids and their derivatives. When cereals are fermented with lactic acid bacteria, for example, the amount of tannin and phytic acid decreases, resulting in increased iron absorption and the elimination of non-nutritive compounds that reduce binding capability. Improve the activity of amylase, phytase, hemicellulose, protease, extend shelf life, digestibility and increased nutritional value (Blandino et al., 2003).

Various types of cereal can be mixed with pulses to make delectable dishes such as idli, dosa, dhokla, ambali and wari. Lactic acid bacteria are thought to be the most common cause of fermentation (LAB). Idli, dosa, adai dosa, kallappam, ambali and dhokla are traditional south Indian foods fermented with *L. mesenteroides, E. faecalis, Pediococcus* sp., *Streptococcus* sp. and L. *fermentum*. The dough is made of basic ingredients and then fermented overnight at room temperature; sometimes sodium bicarbonate is used to provide anaerobic conditions for the growth of LAB (Fig. 5.3) (Satish et al., 2013).

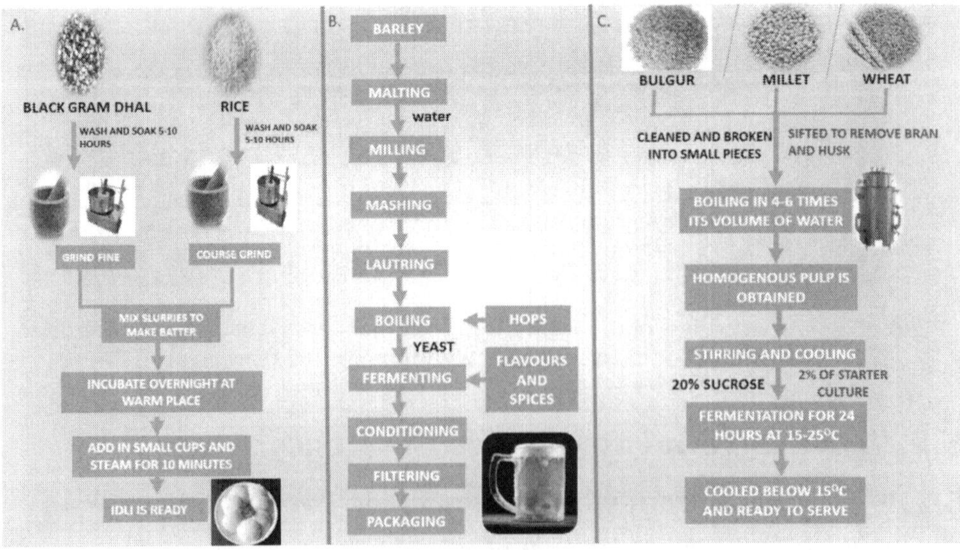

Figure 5.3. - Flow chart for production of grain based fermented food and beverages

A). Idli, a salty cake native to southern India, prepared from the fermented batter. B). Beer, an alcoholic beverage made by the intense fermentation process of Barley. C). Boza, a non-alcoholic Turkish beverage made from the fermentation of the grains.

Modern fermented dishes are mostly made with rice and other cereals. Both of these fermented foods, however, have different quantities and preparation methods depending on where you live (Blandino et al., 2003; Ghosh et al., 2015). The first step is to prepare the dough, which is a thick to semi-solid paste made from raw materials combined with other supplements. This is fermented for a variety of times, either spontaneously or with the aid of a starter. Microbial fermentation affects batter consistency, nutrient pools and food-specific changes (Yousif et al., 2003; Gupta et al., 1992; Chavan et al., 1989).

The fermented batter is then cooked, fried, or baked, depending on the food preparation method. Some foods, on the other hand, are eaten right after they have been fermented. To offer the flavour, fragrance, texture, and taste, the entire process is special and followed dreamily. Fermented foods are usually served with side dishes such as chutneys, curries, snacks, raw vegetables and other condiments (Savitri and Bhalla, 2007)

5.2.1.1 Idli

The salty cake idli in the Indian subcontinent, especially South India is one of the most commonly accepted grain pulse-based naturally fermented foods (Blandino et al., 2003). This rice cake is a low-calorie, potato starch, nutritious breakfast treat. The main ingredients are grain rice (*Oryza sativum*) and legume black gramme dhal. Separately, washed polished rice and black gramme dhal are soaked and ground overnight in drinking water. A dense batter is made by combining rice and black gramme dhal slurries. For fermentation, the batter must be stored at room temperature overnight. The fermented batter is then steamed in a concaved idli tub, resulting in savoury spongy cake idli (Satish et al., 2013). The batter is made using a mixed culture of lactic acid bacteria, including *Leuconostoc mesenteroides, Lactobacillus delbrueckii, Lactobacillus fermenti, Lactobacillus lactis, and Streptococcus faecalis. Saccharomyces cerevisiae, Pediococcus cerevisiae, Debaromyces hansenii, Hansenula anomala, Torulopsis candida* and *Trichosporon beigelii* are among the other species (Satish et al., 2013; Moktan et al., 2011; Purushothaman et al., 1993)

5.2.1.2 Ambali

Ambali is a fermented breakfast food made from rice (*Oryza sativa*) and ragi (*Eleusine coracana*). Ragi flour is ground into a dense slurry and fermented for 14 to 16 hours at room temperature. After that, it's combined with partly cooked rice and cooked some more. To make Ambali, substitute sour milk after it has cooled to room temperature. *Leuconostoc mesenteroids, Lactobacillus fermentum* and *Streptococcus faecalis* are some of the bacteria involved in ambali fermentation

(Blandino et al., 2003). Ambali is a low-calorie fermented food with 115 calories, 1 gramme of total fat, 60 milligrams of sodium, 20 milligrams of total carbohydrate, and 10 milligrams of sugars (approximately 250 millilitres). This is a protein-rich diet that is readily digestible by both children and adults (Ray et al., 2016).

5.2.1.3 Boza

Boza is a beverage made up of millet, corn, semolina or rice & fermented with yeast and lactic acid. It is one of the known non-alcoholic Turkish fermented beverages. It's a viscous, light-yellow liquid with a sweet or sour flavour. It's normally consumed in Turkey, Bulgaria and different Balkan Peninsula countries because of its sensible taste, colour, and organic process properties (Akpinar-Bayizit et al., 2010; Gotcheva et al., 2001; Yegin and Uren, 2008). Lactic acid fermentation, which produces carboxylic acid and makes the drink additional acidic and alcohol fermentation, which produces greenhouse emissions and raises the volume, are each a part of the Boza fermentation method (LeBlanc and Todorov 2011). Lactic acid microorganism conjointly raises the degree of free fatty acids in boza, which provides it with its tasty flavour and prevents unhealthful microorganisms' behaviour (Caputo et al., 2012). In addition to these health benefits, boza is a nourishing and versatile beverage since it contains vitamins, nutrients, carbohydrates, fibre and protein (Todorov 2008, 2010). The two forms of fermentation identified in boza are lactic acid fermentation by LAB and alcohol fermentation by yeasts. Antimicrobial substances are generated by LAB, as well as an increase in acidity, which acts as a preservative. Lactic acid, an antimicrobial substance developed by LAB, is one of the factors that contribute to boza's distinct flavour. Boza's odour and mouthfeel are influenced by alcohol fermentation metabolites. After the fermentation cycle is done, Boza is chilled and distilled in plastic or glass containers (Kabak and Dobson, 2011).

5.2.1.4. Beer

Around 4000 BCE, beer appears to have emerged in ancient Egypt and Mesopotamia (Damerow, 2012). Similar fermented beverages first appeared in Mexico in 2000 BCE and in Sudan in 1500 BCE. (Tamang and Samuel, 2010). Because of its high phenol, vitamin, and melanoidin content, beer has a high antioxidant potential and thus plays an important role in disease prevention (Gonzalez-SanJose et al., 2017). Beer has been brewed since antiquity and is still drunk today with only minor variations of the general formula. The main ingredient is malted barley, which produces a nourishing sugar and protein-rich solution called wort when milled and heated in water to remove its nutrients (pronounced as wert). It's the perfect environment for yeast to thrive and ferment. Hops have been applied to the boiling wort in recent years after it was found that hops had anti-bacterial properties that retained the wort and fermented beer, giving the beer a refreshing bitter flavour (Campbell, 2017). Beer has a distinctive aroma that comes from constituents of malt extracts, hops and yeast metabolism ingredients and has an ethanol content of 2–8% (Tamang and Fleet, 2009). The overall taste & quality of the beer is influenced by the utilization of strains (Dufour et al., 2003). The uncooked materials, mashing, fermentation and conditioning methods, in addition to the yeast type, all play important function in the making of beer (Kurtzman and Robnett, 2003).

5.2.2 Fermented foods/beverages made from milk

Milk is particularly perishable due to the same constituents that make it nutritious. As a result, fermentation is the primary method of preserving milk and its nutrients. (Tamang et al., 2020). Milk, in reality, can be fermented into hundreds of different items. Many forms of lactic acid bacteria (LAB) organisms form spontaneously in milk, resulting in spontaneous fermentation. The growth of LAB causes a sudden drop in pH (acidification), which keeps spoiling microorganisms and pathogens at bay. Bacterial development also aids in the digestion of milk constituents, resulting in increased bioavailability and the formation of flavour compounds.

LAB, which absorbs lactose and produces lactic acid, is involved in the spontaneous fermentation of milk (Carr et al., 2002). *Lactococcus, Lactobacillus, Leuconostoc* and *Pediococcus* are the four most common dairy LAB genera. In addition to producing lactic acid, these bacteria alter other milk constituents, resulting in increased nutritional bioavailability and improved nutrition (Smit et al., 2005). LAB and its metabolic products, in particular, suppress spoilage and pathogenic microorganisms (Arqués et al., 2015). LAB has acidification, which prevents bacteria and microorganisms from multiplying and spoiling food while also releasing antimicrobial bacteriocins (Beermann and Hartung 2013; Widyastuti and Febrisiantosa, 2014).

In general, milk fermentation technology is intuitive and low-cost. Standardized fermented milk products, on the other hand, are processed and manufactured on a wide scale under standardised conditions and have become an important industrial application of LAB as starter cultures. In fermented milk materials, there are a few main characteristics of LAB starts. When comparing fermented milk products to those fermented naturally, a single possible starter culture can control and reduce the variety of microorganisms (Widyastuti and Febrisiantosa, 2014). As a result, fermented dairy foods provide several health advantages, including modulating intestinal microbiota and immune reaction, as well as lowering the risk of hypertension, diabetes and high cholesterol.

One of the most recent thorough studies of the health advantages of fermented dairy products looked at the physiologically active tripeptides valyl-prolyl-proline (VPP) and isoleucyl-prolyl-proline (IPP) (Jauhiainen et al., 2010). These tripeptides are recommended as a biological process technique for delicate high blood pressure (Kim et al., 2010). Eubacteria spp. soured tin can is a potential medical aid for mild hypertension since it produces each ACE-inhibitory peptide and aminoalkanoic acid (g-aminobutyric acid) (Nejati et al., 2013). supported these findings, it's been suggested that carboxylic acid bacteria (LAB) acts as a proteinase on milk proteins throughout fermentation, leading to the event of anti-hypertensive peptides (Hsieh et al.,2015). *Lactobacillus spp*. fortified milk has been shown to help treat hypertension-related cardiovascular disorders (Rodrguez-Figueroa et al., 2013).

5.2.2.1. Dahi

Dahi is a food made from lactic fermentation of cow, buffalo, or mixed milk caused by single or mixed strains of lactic acid bacteria or alcoholic fermentative yeasts, according to the Bureau of Indian Standards (1980) (Deb and Seth, 2014). Dahi is a fermented dairy product that is extremely common in India. It has been used for nutritive and medicinal purposes since time immemorial (Aneja et al., 2002). The use of dahi is prescribed in Ayurveda for the treatment of dysentery, dyspepsia and other gastrointestinal disorders (Gandhi, 2010). Dahi is a lactic acid fermented food made from cow or buffalo milk or a mix of the two. It can be consumed plain or sweetened and salted. It's even eaten with different things like rice and chapatti. Dahi is high in carboxylic acid bacterium and encompasses a probiotic effect, which is useful to viscus health (Sekar and Mariappan, 2007). Mixed strains of lactic acid bacteria and lactose fermenting yeasts are among the microorganisms used in dahi fermentation (Steinkraus et al., 1983; Bhattarai and Lal Das, 2013). Dahi consistency may vary depending on the type of starter culture used and manufacturing techniques used by different manufacturers since standard culture and process are not used when producing dahi. The use of low-quality milk and unsanitary practices during preparation, handling, storage and transportation are all important factors in the development of low-quality dahi (Bhattarai and Lal Das, 2013). In the market, there are various varieties of Dahi, such as soft Dahi, sour Dahi, and flavoured Dahi. Sweet Dahi is commonly made with a mixed culture of S. lactis, S. thermaphilus and S. citrophilus, and sugar is added at a rate of 8-12 percent of the milk amount. The fermented foods have medicinal qualities as well as a high nutritional value. Dahi with a taste is created by adding synthetic flavour or real fruit juice (Sarker, 2018).

5.2.2.2. Cheese

Cheese is a fermented dairy meal of particularly top consistency, with excessive power values and excessive fat, protein, calcium, and diet B content (Ansorena and Astiasaran, 2016). Milk, rennet, starter culture, and proteases and peptidases from secondary microbial plants are used to interrupt down casein and create bioactive compounds chargeable for plenty of organic sports throughout cheese preparation (Lopez-Exposito et al., 2017). The diet and mineral content material of cheese, in addition to bioactive peptides (antihypertensive, antioxidant, opioid, anti-proliferative and antimicrobial peptides and conjugated linoleic acids (CLA)), are usually chargeable for its disease-prevention and remedy impact (Hur et al., 2017). Cheese is assessed as a meal with a whey protein to casein ratio same to or much less than milk. It is made from cultures of good and beneficial microorganisms that produce lactic acid and/or flavour and perfume modifiers and suitable enzymes. Cheese is a microbial fermentation of milk by chosen lactic acid bacteria (LAB), whose primary purpose is to generate lactic acid from lactose, resulting in a pH drop in the curd. It's a diverse microbial world made up of microbes, yeasts and moulds in the case of smear- and mold-ripened cheese. Manufacturing and ripening processes, milk type, water activity, salting methods and the gross chemical composition of the resulting cheeses all have an effect on the microbial flora. Various microorganisms generate enzymes during ripening, especially proteinases and lipases, which hydrolyze protein and fat to amino acids and fatty acids, which are the precursors of the compounds that give the cheese its flavour. Late gas production, also known as "late bursting," occurs late in the ripening process and is caused by *Clostridium tyrobutyricum* and *Clostridium butyricum* fermenting lactate to butyrate, CO_2 and hydrogen. Butyrate causes off-flavouring, while fumes cause big holes in the cheese. Since clostridia can develop during the hot-room ripening time, late gas production can be especially common in Swiss-type cheese (Hayaloglu, 2016).

5.2.2.3. Cultured Buttermilk

Buttermilk is a by-product of the butter production process. Sour-cream buttermilk or sweet-cream buttermilk can be obtained depending on the manufacturing conditions. The microbial fermentation of pasteurised whole milk or skim milk produces Nordic cultured buttermilk. (Tamime et al., 2005). Buttermilk that has been cultured is common fermented milk all over the world. It's made from pasteurised skim or partially skim milk that's been cultured with *Lactococci, Lactobacilli* and aroma-producing strains. During milk fermentation, lactic acid bacteria (LAB) may generate bioactive peptides (Nielsen et al., 2009). Cultured buttermilk prepared from skim milk fermented in a mesophilic environment with lactic and fragrant microorganisms. To raise the temperature of the milk, it is heated to 85°C for 30 minutes. Viscosity while also avoiding wheying-off and killing harmful and pathogenic bacteria. The resulting curd is split and mixed slowly after fermentation, then cooled and lightly salted (Kosikowski and Mistry, 1997). The sour result of fermenting skim or partly skim milk inoculated with LAB culture is cultured buttermilk (Kumar et al., 2015). Buttermilk can help people with moderate hypercholesterolemia lower their blood pressure. Continuous buttermilk intake was found to lower arterial blood pressure and systolic blood pressure. Angiotensin-I converting enzyme levels in plasma have reduced, while angiotensin-II and aldosterone levels remained unchanged (Conwey et al., 2014).

5.2.2.4. Kefir

Kefir is a popular fermented dairy drink with a silky texture and a sour, acidic, and slightly alcoholic flavour (Kabak and Dobson, 2011; Prado et al., 2015; Rai et al., 2017; Sari et al., 2014). It is thought to have originated in the Caucasus. The Caucasus Mountains gave birth to the kefir beverage, which is a common beverage in Eastern Europe, Russia and Southwest Asia. Due to its unusual sensory properties and long history of being linked to positive effects on human health (Farnworth, 2005; Otles and Cagindi, 2003), kefir consumption is currently increasing in many countries. In Brazil, knowledge of this beverage and the benefits of using probiotic foods in a daily diet is limited, and kefir production is primarily artisanal (Carneiro, 2010; Magalhes et al., 2010). Kefir can be made using a variety of techniques, including both conventional and industrial methods. Kefir can be made with any form of milk, including cow, goat and sheep milk, as well as coconut, corn and soy. Pasteurized, unpasteurized, full fat, medium fat, skim, and no fat milk are all options. Similarly, many methods have been developed to make a kefir-like beverage without the use of grains. A mother culture is created in Russia by doing conventional kefir fermentation and sieving the grains. This mother culture is applied to pasteurised milk at a rate of 1 to 3% and incubated for 24 hours at 19 to 28 degrees Celsius. Kefir has been found to have antimicrobial, anti-carcinogenic, probiotic and other functional properties (John and Deeseenthum, 2015) (Fig. 5.4.).

Figure 5.4. - Fermented foods and beverages prepared from milk

5.2.3 Fermented Meat Products

Meat is very prone to microbial spoilage. Meat has traditionally been processed using a variety of methods, including dehydration, curing and fermentation. For a longer preservation time, the North Eastern Indian cultures ferment yak, goat, pig, fish and crab flesh. Natural (uncontrolled) fermentation and controlled fermentation, in which a starting culture is introduced to produce the desired result, are the two types of meat fermentation (Mokoena et al., 2016). Whereas natural fermentation is superior to controlled fermentation in terms of sensory aspects, it permits hazardous bacteria, such as those that produce biogenic amines, to grow (Ojha et al., 2015).

5.2.3.1 Kargyong

It is a traditional fermented sausage made from yak, cattle and pork. Yak kargyong (yak meat), lang kargyong (beef) and faak kargyong (pork) are the three types of kargyong that are cooked and consumed (Satish et al., 2013)

5.2.3.2 Sucuk

It's a famous conventional Turkish fermented meat product, is made with the aid of using cutting red meat or mutton after which including fat, spices, preservatives, colouring materials, additions and starting cultures such as lactic acid bacteria and staphylococci (Akkaya et al., 2014; Eren et al., 2008; Karsloglu et al., 2014).

5.2.3.3 Pastrma

The maximum famous conventional dry-cured semi-fermented meat product, is produced from complete muscle and is extensively ate up and favoured with the aid of use by the Turkish people owing to its sensory qualities (Aksu et al., 2016; Ceylan and Aksu, 2011; Guzin Kaban, 2013). Meat is dried, cured and pressed before being seasoned with cemen, a blend of red pepper, paprika, crushed fenugreek and garlic (Akk€ose et al., 2017; G€ok et al., 2008).

Pastrma was shown to have anti-oxidant and ACE inhibitory qualities in one research, which was linked to the cemen added and the proteolysis process that happened during manufacture (Deniz et al., 2016).

5.3 Conclusion

Almost every village in the globe has its own distinct culinary culture that reflects its ethnic, socioeconomic, and cultural background. Fermented foods and drinks, in particular, have been a key element of human diets for thousands of years on every continent. These ancient foods retain their nutritional and cultural significance in the modern period. Fermented foods and drinks have long been considered to offer health advantages and have been an important component of the human diet. Fermented meals might also additionally decrease the hazard of hypertension, diabetes, obesity, excessive cholesterol, diarrhoea, thrombosis, and different diseases. The bioactive chemicals produced by fermented meals are one explanation for the health advantages they bring. Finally, bioactive molecules enzymes, vitamins, and other compounds produced by the bacteria that cause fermentation have been linked to a variety of health benefits in fermented foods and beverages. However, further research on the degree of intake required to see these health advantages is needed.

References

Akköse, A., Ünal, N., Yalınkılıç, B., Kaban, G. & Kaya, M. (2017). Volatile compounds and some physico-chemical properties of pastırma produced with different nitrate levels. Asian-Australasian journal of animal sciences, 30(8), 1168.

Aksu, M. I., Erdemir, E. & Çakıcı, N. (2016). Changes in the physico-chemical and microbial quality during the production of pastırma cured with different levels of sodium nitrite. Korean journal for food science of animal resources, 36(5), 617.

Akkaya, L., Gök, V., Kara, R. & Yaman, H. (2014). Enterotoxin production by Staphylococcus aureus (A, B, C, D) during the ripening of sucuk (Turkish dry-fermented sausage). CyTA-Journal of Food, 12(2), 127-133.

Ansorena, D. & Astiasarán, I. (2016). Fermented Foods: Composition and Health effects. Encyclopedia of Food and Health. 10.1016/B978-0-12-384947-2.00285-3.

Akpinar-Bayizit, A., Yilmaz-Ersan, L. & Ozcan, T. (2010). Determination of boza's organic acid composition as it is affected by raw material and fermentation. International Journal of Food Properties, 13(3), 648-656.

Aneja, R.P., Mathur, B.N., Chandan, R.C. & Banerjee, A. (2002). Technology of Indian milk products. 1 st edn. Dairy India Publication, New Delhi.

Arqués, J. L., Rodríguez, E., Langa, S., Landete, J. M. & Medina, M. (2015). Antimicrobial activity of lactic acid bacteria in dairy products and gut: Effect on pathogens. Biomedical Research International, 584183. https://doi.org/10.1155/2015/584183

Beermann, C. & Hartung, J. (2013). Physiological properties of milk ingredients released by fermentation. Food & function, 4(2), 185-199.

Bhatia, L. & Paliwal, S. (2010). Banana peel waste as substrate for ethanol production. International Journal of Biotechnology and Bioengineering Research, 1(2), 213–218.

Bhattarai, R.R. & Lal Das, S.K. (2013). Scientific Study on Indigenous Technology of Dahi Making of Eastern Nepal. Journal of Food Processing & Technology, 4, 253.

Blandino, A., Al-Aseeri, M. E., Pandiella, S. S., Cantero, D. & Webb, C. (2003). Cereal-based fermented foods and beverages. Food research international, 36(6), 527-543.

Beltrán-Barrientos, L. M., Hernández-Mendoza, A., Torres-Llanez, M. J., González-Córdova, A. F. & Vallejo-Córdoba, B. (2016). Fermented milk as antihypertensive functional food. Journal of dairy science, 99(6), 4099-4110.

Borgstrom, G. (1968). Principles of food science: Vol. 2 Food microbiology and biochemistry. New York: Macmillan (N.Y.)

Carr, F. J., Chill, D. & Maida, N. (2002). The lactic acid bacteria: a literature survey. Critical reviews in microbiology, 28(4), 281-370.

Cavalieri, D., McGovern, P. E., Hartl, D. L., Mortimer, R. & Polsinelli, M. (2003). Evidence for S. cerevisiae fermentation in ancient wine. Journal of molecular evolution, 57(1), S226-S232.

Chavan, J. K., Kadam, S. S. & Beuchat, L. R. (1989). Nutritional improvement of cereals by sprouting. Critical Reviews in Food Science & Nutrition, 28(5), 401-437.

Caputo, L., Quintieri, L., Baruzzi, F., Borcakli, M. & Morea, M. (2012). Molecular and phenotypic characterization of Pichia fermentans strains found among Boza yeasts. Food Research International, 48(2), 755-762.

Campbell, S. L. (2017). The continuous brewing of beer. DB Breweries Ltd: https://nzic. org. nz/app/uploads/2017/10 A, 6.

Conway, V., Couture, P., Gauthier, S., Pouliot, Y. & Lamarche, B. (2014). Effect of buttermilk consumption on blood pressure in moderately hypercholesterolemic men and women. Nutrition, 30(1), 116-119.

Carneiro, R. P. (2010). Desenvolvimento de uma cultura iniciadora para produção de kefir.

Ceylan, S., & Aksu, M. İ. (2011). Free amino acids profile and quantities of 'sırt', 'bohca' and 'sekerpare' pastirma, dry cured meat products. J Sci Food Agric. Journal of the Science of Food and Agriculture, 91(5), 956-962.

Damerow, P. (2012). Sumerian beer: the origins of brewing technology in ancient Mesopotamia. Cuneiform Digital Library Journal, 2(1).

Deniz, E., Mora, L., Aristoy, M. C., Candoğan, K. & Toldrá, F. (2016). Free amino acids and bioactive peptides profile of Pastırma during its processing. Food Research International, 89, 194-201.

Deb, M. & Seth, D. (2014). Studies on the Physico-Chemical and Microbiological Analysis of Plain Dahi of Assam (India). Journal of Bioresource Engineering and Technology Year, 1, 27-32.

Dufour, J., Verstrepen, K. and Derdelinckx, G. (2003). Brewing yeasts. In Yeast in Foods, eds. T. Boekhout and V. Robert, pp. 347–388. Cambridge, U.K.: Woodhead.

Egounlety, M. & Aworh, O. C. (2003). Effect of soaking, dehulling, cooking and fermentation with Rhizopus oligosporus on the oligosaccharides, trypsin inhibitor, phytic acid and tannins of soybean (Glycine max Merr.), cowpea (Vigna unguiculata L. Walp) and groundbean (Macrotyloma geocarpa Harms). Journal of food engineering, 56(2-3), 249-254.

Eren, I., Yıldız-Turp, G., Kaymak-Ertekin, F. & Serdaroğlu, M. (2008). The effect of external mass transfer resistance during drying of fermented sausage. Drying Technology, 26(12), 1543-1551.

Farnworth, E. R. (2005). Kefir? a complex probiotic. Food Science & Technology Bulletin: Functional Foods, 2, 1-17.

Fruton, J. S. (2006). Fermentation: Vital or chemical process? Leiden: Brill

Gandhi, D.N. (2010). Preparation of a good quality dahi (curd) and probiotic milk products. http://karnaldirectory.com/wp-content/uploads/ 2010 /06/Dahi-Making.pdf.

Ghosh, K., Ray, M., Adak, A., Dey, P., Halder, S. K. & Das, A. (2015). Microbial, saccharifying and antioxidant properties of an Indian rice based fermented beverage. Food Chem, 168, 196-202.

Ghosh, K., Ray, M., Adak, A., Halder, S. K., Das, A., Jana, A. & Mondal, K. C. (2015). Role of probiotic Lactobacillus fermentum KKL1 in the preparation of a rice based fermented beverage. Bioresource technology, 188, 161-168.

González-SanJosé, M. L., Rodríguez, P. M. & Valls-Bellés, V. (2017). Beer and its role in human health. In Fermented foods in health and disease prevention, Academic Press, 365-384.

Gotcheva, V., Pandiella, S. S., Angelov, A., Roshkova, Z. & Webb, C. (2001). Monitoring the fermentation of the traditional Bulgarian beverage boza. International journal of food science & technology, 36(2), 129-134.

Gupta, M., Khetarpaul, N. & Chauhan, B. M. (1992). Preparation nutritional value and acceptability of barley rabadi—an indigenous fermented food of India. Plant foods for human nutrition, 42(4), 351-358.

G€ok, V., Obuz, E. & Akkaya, L. (2008). Effects of packaging method and storage time on the chemical, microbiological, and sensory properties of Turkish pastirma–A dry cured beef product. Meat Science, 80 (2), 335–44. doi:10.1016/j.meatsci.2007.12.017.

Hayaloglu, A. A. (2016). Cheese: Microbiology of cheese. Reference Module in Food Science, 1, 1-11.

Hebert, E. M., Saavedra, L. & Ferranti, P. (2010). Bioactive peptides derived from casein and whey proteins (p. 233). Ames: Wiley-Blackwell.

Hesseltine. C.W. & Wang, H.L. (1980). The importance of traditional fermented foods. BioScience. Food from Microbes. 30(6), 402-404.

Hutkins, R. W. (2007). Microbiology and Technology of Fermented Foods. Blackwell Publishing. https://doi.org/10.1002/9780470277515

Hur, S. J., Kim, H. S., Bahk, Y. Y. & Park, Y. (2017). Overview of conjugated linoleic acid formation and accumulation in animal products. Livestock Science, 195, 105-111.

Jauhiainen, T., Pilvi, T., Cheng, Z. J., Kautiainen, H., Müller, D. N., Vapaatalo, H., Korpela, R. & Mervaala, E. (2010). Milk Products Containing Bioactive Tripeptides Have an Antihypertensive Effect in Double Transgenic Rats (dTGR) Harbouring Human Renin and Human Angiotensinogen Genes. Journal of nutrition and metabolism. https://doi.org/10.1155/2010/287030

Jeyaram, K. (2009). Traditional fermented foods of Manipur. Indian Journal of Traditional Knowledge, 8 (1), 115-121.

John, S. M. & Deeseenthum, S. (2015). Properties and benefits of kefir-A review. Songklanakarin Journal of Science and Technology, 37(3), 275-282.

Kabak, B. & Dobson, A. D. (2011). An introduction to the traditional fermented foods and beverages of Turkey. Critical reviews in food science and nutrition, 51(3), 248-260.

Karsloğlu, B., Çiçek, Ü. E., Kolsarici, N. & Candoğan, K. (2014). Lipolytic changes in fermented sausages produced with turkey meat: effects of starter culture and heat treatment. Korean journal for food science of animal resources, 34(1), 40.

Kim, S. M., Park, S. & Choue, R. (2010). Effects of fermented milk peptides supplement on blood pressure and vascular function in spontaneously hypertensive rats. Food Science and Biotechnology, 19(5), 1409-1413.

Kurtzman, C. P. & Robnett, C. J. (2003). Phylogenetic relationships among yeasts of the 'Saccharomyces complex' determined from multigene sequence analyses. FEMS yeast research, 3(4), 417-432.

Kosikowski, F. V. & Mistry, V.V. (1997). Cheese and Fermented Milk Foods, 3rd edn., ed. F. V. Kosikowski. Westport, CT: LLC

Leblanc, J. G.J., & Todorov, S.D. (2011). Bacteriocins producing lactic acid bacteria isolated from Boza, a traditional fermented beverage from Balkan Peninsula–from isolation to application.

López-Expósito, I., Miralles, B., Amigo, L. & Hernández-Ledesma, B. (2017). Health effects of cheese components with a focus on bioactive peptides. In Fermented foods in health and disease prevention, Academic Press, 239-273.

Magalhaes, K. T., Pereira, G. D. M., Dias, D. R. & Schwan, R. F. (2010). Microbial communities and chemical changes during fermentation of sugary Brazilian kefir. World Journal of Microbiology and Biotechnology, 26(7), 1241-1250.

Martinez-Villaluenga, C., Penas, E. & Frias, J. (2017). Bioactive peptides in fermented foods: Production and evidence for health effects. In Fermented foods in health and disease prevention, Academic Press, 23-47.

Marco, M. L., Heeney, D., Binda, S., Cifelli, C. J., Cotter, P. D., Foligné, B. ... & Hutkins, R. (2017). Health benefits of fermented foods: microbiota and beyond. Current opinion in biotechnology, 44, 94-102.

McGovern, P. E., Zhang, J., Tang, J., Zhang, Z., Hall, G. R., Moreau, R. A. ... & Wang, C. (2004). Fermented beverages of pre-and proto-historic China. Proceedings of the National Academy of Sciences, 101(51), 17593-17598.

Moktan, B., Roy, A. & Sarkar, P. K. (2011). Antioxidant activities of cereal-legume mixed batters as influenced by process parameters during preparation of dhokla and idli, traditional steamed pancakes. International journal of food sciences and nutrition, 62(4), 360-369.

Mokoena, M. P., Mutanda, T. & Olaniran, A. O. (2016). Perspectives on the probiotic potential of lactic acid bacteria from African traditional fermented foods and beverages. Food & Nutrition Research, 60(1), 29630.

Nejati, F., Rizzello, C. G., Di Cagno, R., Sheikh-Zeinoddin, M., Diviccaro, A., Minervini, F. & Gobbetti, M. (2013). Manufacture of a functional fermented milk enriched of Angiotensin-I Converting Enzyme (ACE)-inhibitory peptides and γ-amino butyric acid (GABA). LWT-Food Science and Technology, 51(1), 183-189.

Nielsen, C. B., Jackman, S. D., Birol, I. & Jones, S. J. (2009). ABySS-Explorer: visualizing genome sequence assemblies. IEEE transactions on visualization and computer graphics, 15(6), 881-888.

Ojha, K. S., Kerry, J. P., Duffy, G., Beresford, T. & Tiwari, B. K. (2015). Technological advances for enhancing quality and safety of fermented meat products. Trends in Food Science & Technology, 44(1), 105-116.

Otağ, F. B. & Hayta, M. (2013). Gıdalarda Biyoaktif Peptit Oluşumu ve Aktivitesi Üzerine Isıl İşlem Ve Fermantasyonun Etkileri. Gıda, 38(5), 307-314.

Otles, S. & Cagindi, O. (2003). Kefir: A probiotic dairy-composition, nutritional and therapeutic aspects. Pakistan journal of nutrition, 2(2), 54-59.

Ozdemir, S., Gocmen, D. & Yildirim Kumral, A. (2007). A traditional Turkish fermented cereal food: Tarhana. Food Reviews International, 23(2), 107-121.

Prado, M. R., Blandón, L. M., Vandenberghe, L. P., Rodrigues, C., Castro, G. R., Thomaz-Soccol, V. & Soccol, C. R. (2015). Milk kefir: composition, microbial cultures, biological activities, and related products. Frontiers in microbiology, 6, 1177.

Purushothaman, D., Dhanapal, N. & Rangaswami, G. (1993). Indian idli, dosa, dhokla, khaman, and related fermentations. Handbook of indigenous fermented foods. Marcel Dekker, New York, 149-165.

Rai, A. K., Sanjukta, S. & Jeyaram, K. (2017). Production of angiotensin I converting enzyme inhibitory (ACE-I) peptides during milk fermentation and their role in reducing hypertension. Critical reviews in food science and nutrition, 57(13), 2789-2800.

Ray, M., Ghosh, K., Singh, S. & Mondal, K. C. (2016). Folk to functional: an explorative overview of rice-based fermented foods and beverages in India. Journal of Ethnic Foods, 3(1), 5-18.

Rezac, S., Kok, C. R., Heermann, M. & Hutkins, R. (2018). Fermented foods as a dietary source of live organisms. Frontiers in microbiology, 9, 1785.

Rodríguez-Figueroa, J. C., González-Córdova, A. F., Astiazaran-García, H. & Vallejo-Cordoba, B. (2013). Hypotensive and heart rate-lowering effects in rats receiving milk fermented by specific Lactococcus lactis strains. British Journal of Nutrition, 109(5), 827-833.

Sari, E. K., Bakir, B., Aydin, B. D. & Sozmen, M. (2014). The effects of kefir, koumiss, yogurt and commercial probiotic formulations on PPARα and PPAR-β/δ expressions in mouse kidney. Biotechnic & Histochemistry, 89(4), 287-295.

Sarker, M. T., Prabakusuma, A.S. & Islam, M.S. (2018). Dahi (Curd) preparation from milk with different levels of carrot (Dacus carota) juice. MOJ Food Process Technol, 6(1), 66-71.

Satish, K.R., Kanmani, P., Yuvaraj, N., Paari, K. A., Pattukumar, V. & Arul, V. (2013). Traditional Indian fermented foods: a rich source of lactic acid bacteria. International journal of food sciences and nutrition, 64(4), 415-428.

Sekar, S. & Mariappan, S. (2007). Usage of traditional fermented products by Indian rural folks and IPR.

Savitri, T.A. & Bhalla, T.C. (2007). Traditional foods and beverages of Himachal Pradesh. Indian J Tradit Know. 6, 17-24.

Savage, D. C. (1986). Gastrointestinal microflora in mammalian nutrition. Annual review of nutrition, 6(1), 155-178.

Sekar, S. & Mariappan, S. (2007). Usage of traditional fermented products by Indian rural folks and IPR.

Slashinski, M. J., McCurdy, S. A., Achenbaum, L. S., Whitney, S. N. & McGuire, A. L. (2012). "Snake-oil" quack medicine," and "industrially cultured organisms:" biovalue and the commercialization of human microbiome research. BMC medical ethics, 13(1), 1-8.

Smit, G., Smit, B. A. & Engels, W. J. (2005). Flavour formation by lactic acid bacteria and biochemical flavour profiling of cheese products. FEMS microbiology reviews, 29(3), 591-610.

Steinkraus, K. H., Cullen, R. E., Pearson, C. S., Nellis, L. E. & Gavitt, B. K. (1983). Handbook of Indigenous Fermented Foods. Marced Dekker. Inc., New York.

Steinkraus, K. H. (1994). Nutritional significance of fermented foods. Food Research International, 27, 259–267.

Tamang, J.P. & Kailasapathy, K. (Eds.). (2010). Fermented Foods and Beverages of the World (1st ed.). CRC Press. https://doi.org/10.1201/EBK1420094954

Tamang, J. P., Shin, D.-H., Jung, S.-J. & Chae, S.-W. (2016). Functional properties of microorganisms in fermented foods. Frontiers in Microbiology. doi:10.3389/fmicb.2016.00578.

Tamang, J. P., Thapa, N., Tamang, B., Rai, A. & Chettri, R. (2015). Microorganisms in fermented foods and beverages. Health benefits of fermented foods and beverages, 1-110.

Tamang, J. P., Watanabe, K., & Holzapfel, W. H. (2016). Diversity of microorganisms in global fermented foods and beverages. Frontiers in microbiology, 7, 377.

Tamang, J. P., Cotter, P. D., Endo, A., Han, N. S., Kort, R., Liu, S. Q. & Hutkins, R. (2020). Fermented foods in a global age: East meets West. Comprehensive Reviews in Food Science and Food Safety, 19(1), 184-217.

Tamang, J. P. & Samuel, D. (2010). Dietary culture and antiquity of fermented foods and beverages. In J. P. Tamang & K. Kailasapathy (Eds.), Fermented foods and beverages of the world (pp. 1–40). New York, NY: CRC Press

Tamang, J. P. & Fleet, G. H. (2009). Yeasts diversity in fermented foods and beverages. In Yeast biotechnology: diversity and applications. Springer, Dordrecht, 169-198.

Tamime, A. Y., Saarela, M. A. K. S., Sondergaard, A. K., Mistry, V. V. & Shah, N. P. (2005). Production and maintenance of viability of probiotic microorganisms in dairy products. Probiotic dairy products, 3, 39-63.

Todorov, S. D. (2008). Bacteriocin production by Lactobacillus plantarum AMA-K isolated from Amasi, a Zimbabwean fermented milk product and study of the adsorption of bacteriocin AMA-K to Listeria sp. Brazilian Journal of microbiology, 39, 178-187.

Todorov, S. D. (2010). Diversity of bacteriocinogenic lactic acid bacteria isolated from boza, a cereal-based fermented beverage from Bulgaria. Food Control, 21(7), 1011-1021.

Usinger, L., Ibsen, H. & Jensen, L. T. (2009). Does fermented milk possess antihypertensive effect in humans? Journal of hypertension, 27(6), 1115-1120.

Vouillamoz, J. F., McGovern, P. E., Ergul, A., Söylemezoğlu, G., Tevzadze, G., Meredith, C. P. & Grando, M. S. (2006). Genetic characterization and relationships of traditional grape cultivars from Transcaucasia and Anatolia. Plant genetic resources, 4(2), 144-158.

Widyastuti, Y., Rohmatussoliha & dan Febrisiantosa, A. (2014). The Role of Lactic Acid Bacteria in Milk Fermentation. Food and Nutrition Sciences, 5, 435-442.

Walther, B. & Sieber, R. (2011). Bioactive proteins and peptides in foods. International Journal for Vitamin and Nutrition Research, 81(2), 181.

Yeğin, S. & Üren, A. (2008). Biogenic amine content of boza: A traditional cereal-based, fermented Turkish beverage. Food Chemistry, 111(4), 983-987.

Yousif, N. E. & El Tinay, A. H. (2003). Effect of natural fermentation on protein fractions and in vitro protein digestibility of rice. Plant Foods for Human Nutrition, 58(3), 1-8.

Zhu, Y. & Tramper, J. (2013). Koji-where East meets West in fermentation. Biotechnol Adv, 31, 1448-57.

Chapter-6

DOWNSTREAM PROCESSING

Dowluru SVGK Kaladhar [1]* ***and Srinivasan Tantravahi*** [2]

[1]*Department of Microbiology and Bioinformatics, UTD, Atal Bihari Vajpayee University, Bilaspur (CG), India*

[2]*Department of Botany, IGNTU, Amarkantak (MP), India*

**Corresponding author Email: dkaladhar@gmail.com*

Abstract

Downstream process refers to technology for separation and purification of biochemical's that is used to be employed in biotechnology industry. The term 'biologics' are used for several therapeutic proteins that are used as engineered macromolecular products. continuous downstream processing in industries showing high value biological products is very much essential for processing of therapeutic proteins. The process of separation, purification and recovery of desired products using various suitable techniques and methods for recovery take place after the fermentation or bioconversion stage. Most of the recovery applies filtration and chromatography in several biopharmaceutical industries. Industrial products like oil, cell culture, proteins and enzymes, amino acids, vitamins, monoclonal antibodies, antibiotics, alcohols, algal products, gasoline, asphalt, diesel fuel, cGMP products are concerned *with the extraction* in downstream processing. The *challenges and benefits for the development of several* downstream processes require *rapid, simple and inexpensive methods. The present information provides current methods in downstream processing that shows different techniques behind the process for product isolation, purification and delivery. Conventional downstream methods generally cover methods like filtration, centrifugation, sedimentation, flotation, precipitation, electro precipitation, gravity settling, grinding, high pressure homogenization, etc.*

Keywords: *Downstream processing, Fermentation, desired products.*

6.1 Introduction

Fermentation is a microbial process that relates to generation of the energy by the catabolism process of the organic compounds (Kleerebezem and van Loosdrecht, 2007; Kaladhar, 2014). The term fermentation was derived from a Latin word "Fevere" which means "Boil" (Ditu and Gheorghe, 2017; Tsafrakidou et al., 2020). Products like Microbial alkaline proteases from *Bacillus* sp. using chromatographic and crystallization (Gupta et al., 2002), algal polyunsaturated fatty acids from microalgae using urea fractionation and high-performance liquid chromatography (Medina et al., 1998), xylanases from *Trichoderma sp.* using column chromatographic techniques, ion exchange and gel filtration (Bajpai, 2014) are some of the downstream processes.

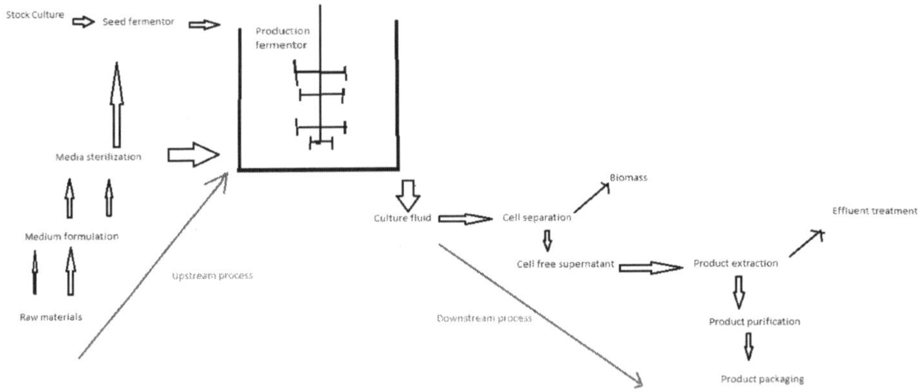

Figure 6.1. - Schematic diagram of a typical fermentation process

The downstream process is divided into six basic components like microbial culturing, sterilization, production of pure culture, production in fermentor, extraction and purification, and disposal of effluents (Anastassiadis, 2007; Sankaran et al., 2010) There are different types of products in the process of fermentation in industrial products contain 3 processes: upstream process, fermentation and downstream processing (Fig. 6.1). The form of products is of 3 types: a cell like yeast, extracellular type and intracellular types (Brauchle and Schenke-Layland, 2013).

Manufacturing of several types of Biotechnology products were gradually evolved from several years using traditional batch mode operation in the production of alcohols to continuous modes of operation process in the modern robotic industries. The technical advancements in the development of the continuous chromatography focus have improved modernization in the industries to an integrated upstream and downstream continuous process (Lute et al., 2020). Downstream processes are influenced by product concentration and stability (Fig. 6.2).

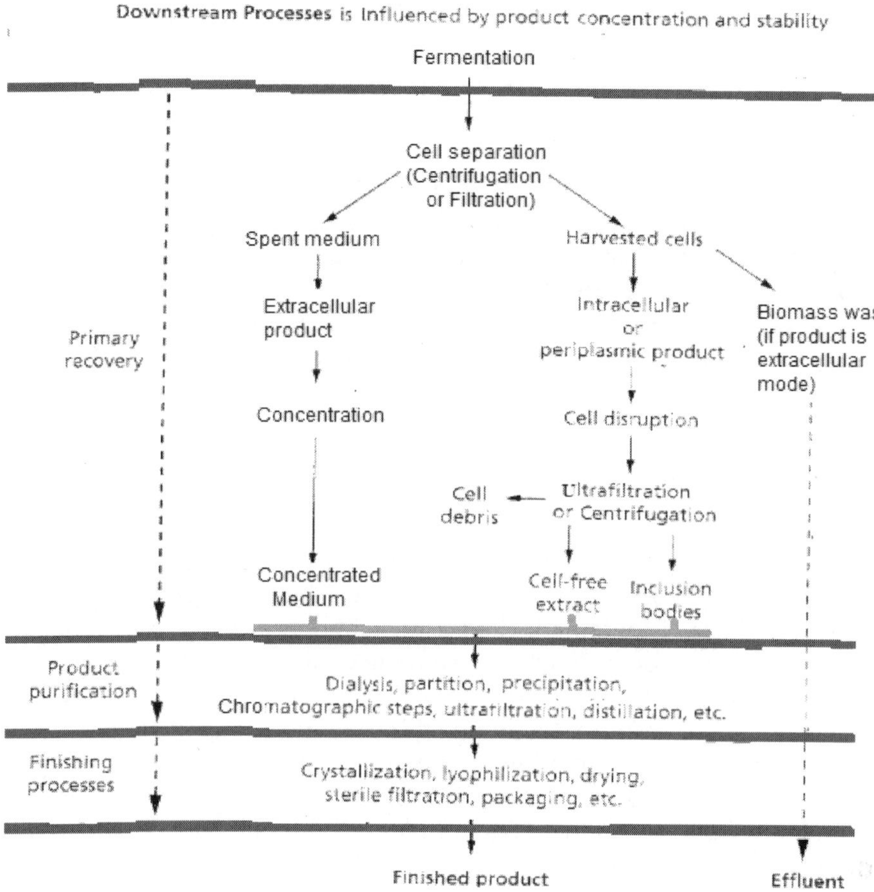

Figure 6.2. - Downstream processing

Biological products for commercial applications contain several sources like plant material (oils, Taxol from the bark of Taxus species), microbial fermentations, human and animal tissue (blood, pituitary, pancreas), body fluids (milk of transgenics), cultures of higher eukaryotes (Chisti, 1998). Downstream processing refers to the process of extraction and purification of the biological products after fermentation especially by using microorganisms. Downstream is generally very expensive process due to instrumentation usage in biochemical or molecular biology techniques for microbial fermentation (Wang et al., 2019). The present information provides current methods in downstream processing that shows different techniques behind the process for product isolation, purification and delivery. Conventional downstream methods generally cover methods like filtration, centrifugation, sedimentation, flotation, precipitation, electro precipitation, gravity settling, grinding, high pressure homogenization, etc.

6.2. Removal of Insoluble Products

The first phase in recovering a product is to separate the cells and other insoluble substances from the broth (Lee et al., 2019; Yesil-Celiktas and Vardar-Sukan 2013). If the target product is an intracellular metabolite, the cells must first be extracted. Cellular components, which are influenced by the host cell composition, are the most significant contaminants. The most effective method for softening the biomass, residual substrate, starch-based sizes, cell debris and small particles will be conducted in the first phase of downstream processing (Bergemann et al., 2007; Bhatia et al., 2018). Impurities can, however, be detected in secreted products since cell lysis occurs at all stages of the fermentation process. The removal of insoluble process during fermentation is also known as 'cell harvesting' (Bals et al., 2010).

Filtration

The most conventional method of separating biomass from culture filtrate is filtration (Taniguchi et al., 1987; Manikandan et al., 2019). The liquid component of the broth is separated from the solid component by passing the fluid-solid mixture through filter. The solid components including the biomass are retained on the filter and clear filtrate is obtained. The filtration and its efficiency are reduced by the gradual accumulation of filter cake (Hwang et al., 2009; Lower et al., 2020). To overcome this, the positive pressure equipment or vacuum is employed. It is essential to note that several parameters influence filtration efficiency, including organism size, the presence of other organisms in the filtering system, medium viscosity, and temperature. Pressure is determined by the filter area, fluid viscosity, and filter bed resistance, and it rises over time. There are three major types of filtrations based on particle size: microfiltration, ultrafiltration, and reverse osmosis. Microfiltration separates particles with diameters ranging from 0.1 to 10μm, the majority of which are cells and their fractions (Bowen and Jenner, 1995; Saboyainsta and Maubois, 2000). Ultrafiltration separates particles with diameters of 0.001-0.01 μm, such as proteins and enzymes, whereas Hyperfiltration (or reverse osmosis) separates particles with diameters of 0.0001-0.001 μm, such as various chemical compounds. Microfiltration is often utilized to separate entire cells from broth and the collection of protein precipitates and inclusion bodies (Table 6.1, Fig. 6.3).

Table 6.1 - Types of filtrations

Type	*Size of particles to separate*	*Samples in this process*
Reverse Osmosis (RO)	-0.0005 μm to 5Å	Chloride ions and atomic radii
Nanofiltration (NF)	-0.001 μm to 10Å	Sugars, dyes, metal ions, salts, alcohols
Hyperfiltration (HF)	0.0001 to 0.001μm	Compounds with very less molecular weight <1000 amu (Example chemical compounds like lactose)

Type	Size of particles to separate	Samples in this process
Ultrafiltration (UF)	0.001 to 0.1 μm	Compounds with very less molecular weight >1000amu (Example Proteins and Enzymes)
Microfiltration (MF)	0.1 to 10 μm	Cell fractions and viruses
Conventional filtration (CF)	-1 μm to 100000 Å	Fiber, sand, hair, RBC etc

The extraction of micron-sized paramagnetic particles can be constructed by passing fluids through the magnetic filters. The main application of magnetic filtering of the product is used in the rapid treatment of sewage in the industries (Watson, 1973). Nano filtration is an effective process of size-based mechanism in the modern times that is applied for removal of virus particles. Stirred well filtration (SWF) is a high-throughput filtration technique showing hydrodynamic and protein sieving process that applies transmembrane pressure (TMP) on stirred well filter plate (Kazemi and Latulippe, 2014). Hence downstream operations in the biopharmaceutical productions have been enabled with more flexible manufacturing operations using continuous flow filtration platform processes (Brinkmann et al., 2018).

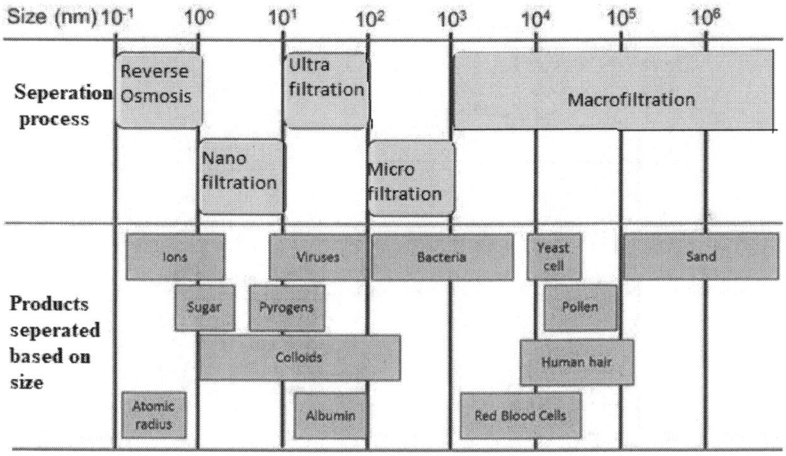

Figure 6.3. - Filtration systems

Centrifugation

Centrifugation is the approach that is used to separate cells, proteins, and debris based on density differences by applying centrifugal forces that complement the gravity (Leung, 2020). Centrifugation has the advantage of making it easier to separate the entire biomass from the fermentation broth. Concentrated suspensions are effectively separated over the dilute suspensions (Grima et al., 2004). Continuous centrifuges are exclusively employed in large-scale production to recover precipitated proteins or contaminants (Hammerschmidt et al., 2016).

Particles with small variations in density cannot be separated by centrifugation. Furthermore, only a limited volume can be handled simultaneously, and being an open process will lead to contamination. Centrifugation is a reasonably uncommon technique to employ based on its operational costs. The limits of the method are increased by the expensive cost of equipment, high power consumption, and other factors.

Several challenges in downstream processing for biologics through primary recovery and purification Magnetically Enhanced Centrifugation (MEC) uses synthetic magnetic particles that transport out the magnetic field by simulations in industrial large-scale machines (Lindner, 2014). Subcellular fractionation process is an enrichment technique that allows the identification of proteins and organelle constituents by isolation and endosomal purification methods based on differential centrifugation (Araujo et al., 2008). From the final products, the removal of bacterial and viral contaminants for production of high-value therapeutic proteins from mammalian cells, recombinant bacteria, and transgenic animals can be achieved by centrifugation and membrane filtration (Kalyanpur, 2002). The are several types of centrifuges like Scroll or Decanter centrifuge, Tubular bowl centrifuge, Disc centrifuge, Disc-stack centrifuge, multi-chamber centrifuge, etc. (Fig. 6.4).

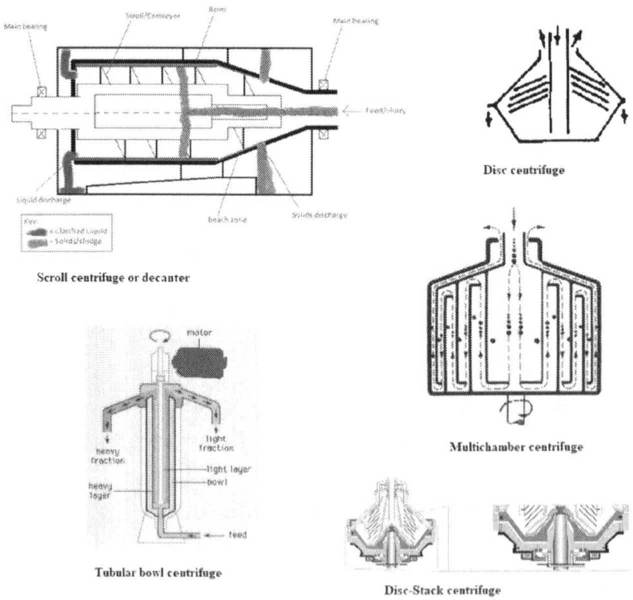

Figure 6.4. - Some types of centrifuges

Sedimentation

Sedimentation is a simple process in which big particles larger than 100 μm, such as biomass or secreted proteins, settle at the vessel's bottom. a high amount of particles in the sediment greater than 600 μm (Leung, 2020; Nitsos et al., 2020)). It is a slow process and works by gravitational force. The sedimentation technique is also a part of the flocculation process where the solute is separated from the solution. The polyvalent ions or polymers are also employed to make the cell aggregate. Commercial production of biomolecules such as carbohydrate (carageenan, agar, alginate, fucoidan), pigments (Xanthophyll, chlorophyll and carotenoid), biopolymers and protein from microalgae and cyanobacterial cells can be achieved in downstream processes through sedimentation process (Khanra et al., 2018). The morphologically engineering of cells from small microorganisms can be induced aggregated to spheres, so that press filtration or gravity sedimentation can becomes a convenient operation for easy cell separation in the downstream processing (Wang et al., 2019).

Particles in the fluid suspension settle out in the sedimentation process which is very effective in removing the fine particles (Fig. 6.5). Sedimentation tanks may be of different types like circular with radial flow, rectangular with horizontal flow, or hopper-bottomed with upward flow (Scholz, 2015).

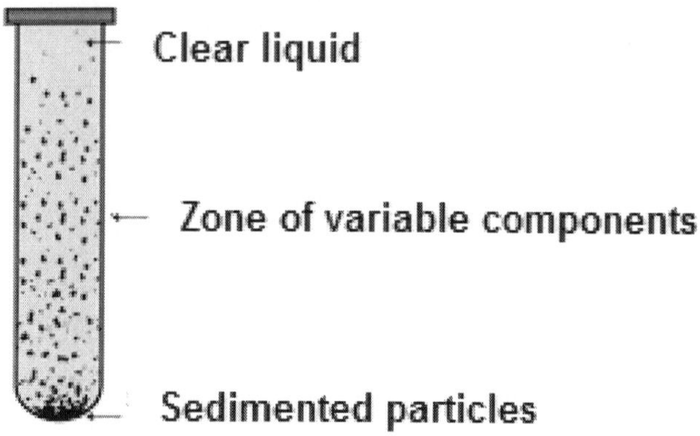

Figure 6.5. - Sedimentation

Flocculation

Flocculation technique is used in downstream processes where cells or secreted products are released as flakes or flocs using polyvalent ions or extracellular polymers (Branyikova et al., 2018). The broth is sedimented after flocculation to separate it. Product flocculation can aid easy isolation and purification, whereas fermenting organism flocculation can help reuse and product purification. Alum, lime, and other natural and chemical coagulants are employed in the process. After completing an enzyme synthesis reaction, adding organism-specific flocculants to the clear solution can settle the organisms to the bottom, isolating the enzyme. A homopolymer composed of glutamic acid residues from biopolymer flocculant-producing bacterium *Bacillus subtilis* strain PY-90 can be achieved in downstream processing (Yokoi et al., 1995). During the flocculation process, small particles will form large aggregates in the solution that improves settling rates. In some previous studies, microalgal C. vulgaris biomass *flocculation* by chitosan has been reported during the harvesting affect of downstream lipid extraction (Fig. 6.6).

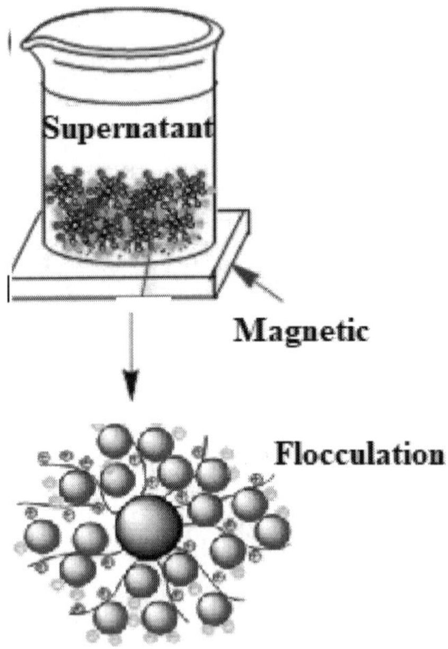

Figure 6.6. - Flocculation

Precipitation

Precipitation process differs from flocculation, where precipitation process is frequently used to separate proteins from broth. Precipitation method is ineffective for dilute broths. Protein precipitation is commonly accomplished with ammonium sulphate, whereas polysaccharide precipitation is accomplished with alcohol. Ammonium sulphate is an inorganic salt that is used at high ionic strength and is popular due to its low toxicity, high solubility and low cost. The procedure also makes use of acetone, polyethylene glycol, polypropylene glycol, and other compounds. Organic solvents have the drawback of generating a reducing environment, which can be mitigated by processing at low temperatures. Another approach for precipitating a material is to change the pH or the temperature. Precipitation reactions are carried out in batch and continuous reactors.

Broad molecular variety of new products like plasmid DNA, virus-like particles, exosomes and monoclonal antibodies (mAb) can be achieved by process integration of precipitation in downstream processing (Lohmann and Strube, 2020). Several fermentation products like carbohydrates can be precipitated by the addition of alcohol and organic acids or proteins are been salted out of the solution as precipitates by centrifugation (Fig. 6.7).

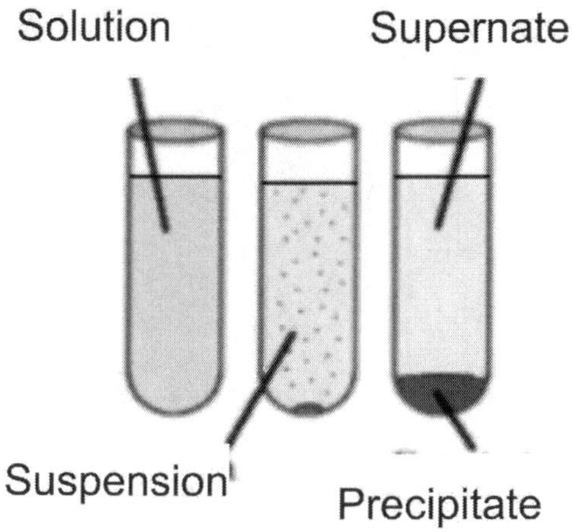

Figure 6.7. - Precipitation

Electro-precipitation

Electro-precipitation method separates microorganisms (Kliaugaitė et al., 2013) products, and byproducts from the broth using an electric current. Because of their ionic connections, the molecules may dissolve in the broth. The flow of electric current will destabilize the interactions by changing the existing individual charges. As a result, the molecules congregate and precipitate (Musa et al., 2020). Selective precipitation of the compounds can be achieved by adjusting the voltage and period of current flow. The precipitates are separated and processed further until a pure compound is obtained. Electro-precipitation method is also useful for separating molecules from a gaseous mixture.

Separating the cells and other insoluble material from the broth is the first step in recovering a product (Hao et al., 2006; Dai et al., 2014). If the target product is an intracellular metabolite, the cells must first be removed. The most significant contaminants are cellular components, which are impacted by the nature of the host cell. Impurities can, however, be detected in released products since cell lysis occurs at every stage of the fermentation process. This procedure is also known as cell harvesting.

The four basic types of Electrostatic Precipitators (ESPs) are flat plate (dry), flat plate (wet), plate and wire (dry), and two-stage. Figure 6.8 shows electro-precipitator that applies in **Electro-precipitation** method during downstream process.

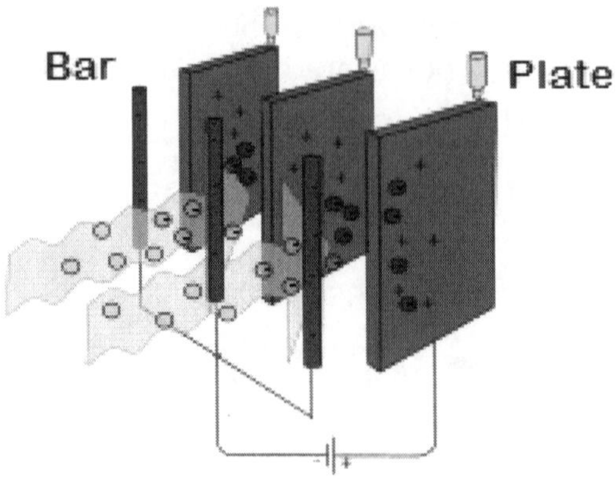

Figure 6.8. - Electro-precipitator

Gravity Settling

Particles subjected to a force, such as gravity or centrifugal motion, will tend to travel uniformly in the force's direction (Caulet et al., 1996). Gravity settling happens when particles fall to the vessel's bottom, generating a slurry. Gravity settling takes a long time for larger and heavier particles. Because most microorganisms are small and light, they are programmed to develop into larger and heavier ones that are subsequently sedimented by gravity (Fig. 6.9). Gravity settling is utilized in a variety of applications, such as large-scale downstream processing of metabolites or products, wastewater treatment, and so on.

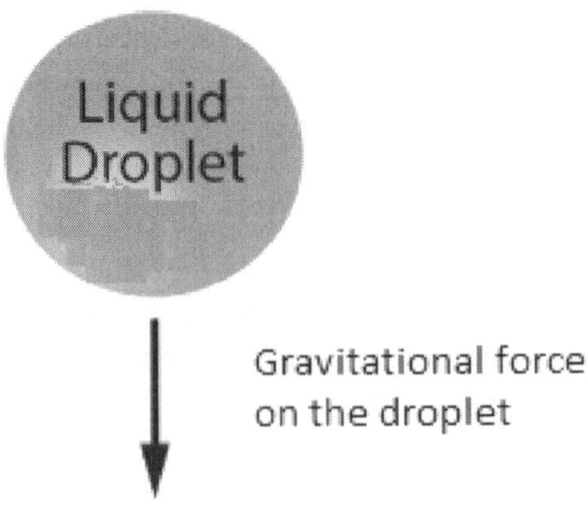

Figure 6.9. - Gravitational force

Gravitational settling is one of the most efficient means of separating dense solids from surrounding liquid with a certain viscosity (Ma and Hanna, 1999). The benefits of gravitational settling are low maintenance and energy cost, disposal of dry and continuous solid particulates, and low pressure drop through the devices.

Grinding

Various grinding processes are used in the early stages of downstream processing to release products from cells or tissues into the broth. Tools that exert mechanical pressure, such as a mortar and pestle or glass beads, are used to disturb cells in order to liberate intracellular contents or products (Lee et al., 2019). Grinding or milling as a significant impact on product yield because it is the first downstream process. The higher the efficiency of the process, the larger will be the yield. Numerous parameters, including cell concentration and age, equipment temperature and speed, and bead size and quantity, continue to influence cell lysis effectiveness.

An effective means of disruption of cells using abrasives like usually glass beads, alumina, silica or kieselguhr is effective in manual grinding (Burden, 2012). The algal cysts contain carotenoid will be isolated and extracted by *grinding* the dried samples of algae. Cell breakage depends due to factors like concentration of beads, agitation speed, broth density, temperature, bead density and diameter and flow rate during grinding. The mortar and pestle that use in grinding process has been shown in figure 6.10.

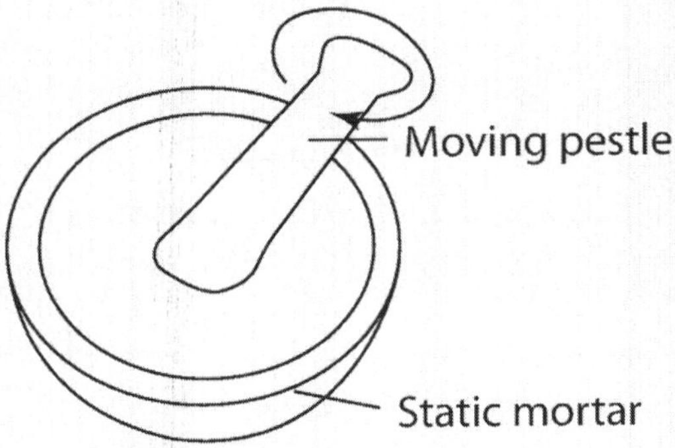

Figure 6.10. - Mortar and pestle

High Pressure Homogenization

The cells are homogenised by lysing them under high pressure. This process is frequently utilized when grinding results are ineffectual, as in yeast, fungus, and other organisms with cell walls that are resistant to cell lysis (Diels and Michiels, 2006). High pressure homogenization method includes introducing a high-pressure cell suspension through a tiny orifice into atmospheric pressure. Cells burst as a result of an abrupt change in pressure (Fig. 6.11 and 6.12). Homogenizers, which can break cell walls of varying strengths without causing damage to intracellular components, are already available.

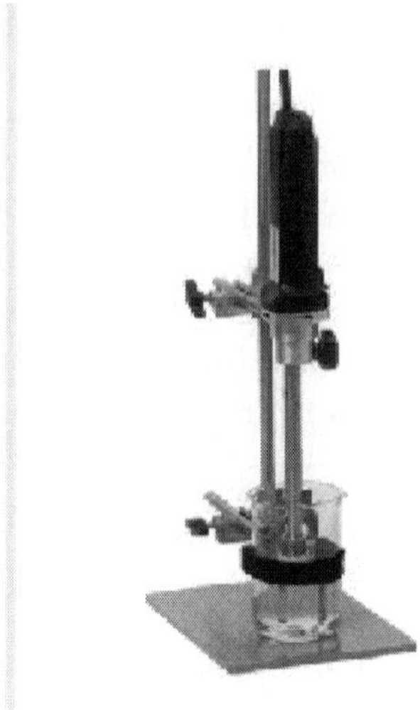

Figure 6.11. - Homogenizer

Establishing high shear stress used to push liquids through a narrow gap (micron size) at high pressures between 15 and 2000 bars on cells is used to produce high-quality products. Shear-sensitive products like DNA is not typically recovered through mechanical approaches.

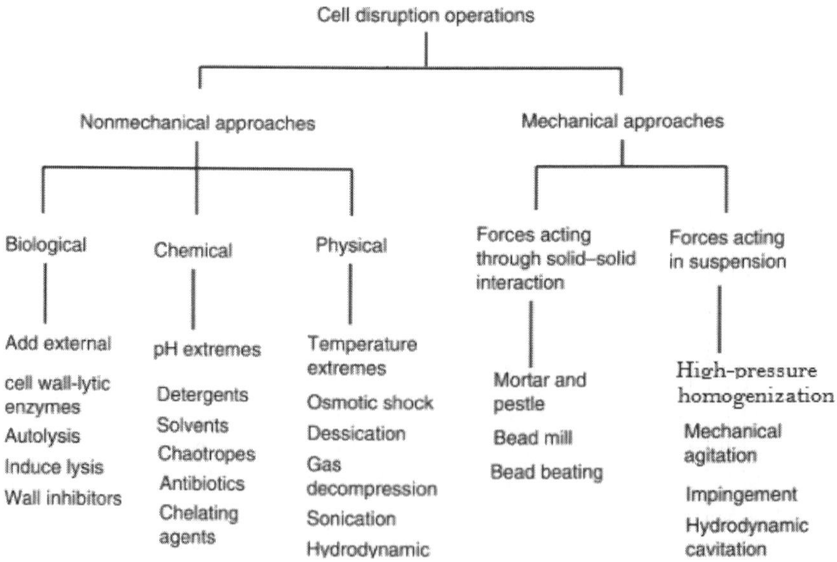

Figure 6.12. - Cell disruption operations

6.3 Product Isolation

The product must be purified after cell lysis and/or separation of the product from the broth. In this stage of downstream processing, compounds that differ significantly from the compound of interest are separated (Kalyanpur, 2002). Isolation methods are designed in accordance with the nature of the product and/or phase of the product in which it is present.

Solid particle separation is influenced by the particle size, viscosity, density, and foam formation in the broth/liquid component. The separation of liquid products depends on their miscibility. The immiscible products can be separated based on the phase they are in. Miscible products necessitate a variety of procedures for isolation and purification. Finally, the desired product concentration increases considerably and substances of wide polarities with that of the product are eliminated.

The recovery and purification through extraction, adsorption, precipitation or ultrafiltration is the significant challenges over the past few years during product isolation (Galanakis, 2013). The removal of the desired products like hormones (e.g., insulin and human growth hormone), antibiotics, vaccines, alcohols, antibodies, natural fragrance and flavor compounds, organic acids, vitamins, enzymes, etc., can be achieved in the product isolation process.

Solvent Extraction

Antibiotics, vitamins, lipids, and other compounds are highly soluble in organic solvents when compared to water (Didaras et al., 2020). A multistage solvent extraction separates diverse organic phases and yields both a concentrated and purified product. The solvents used should be nontoxic, inexpensive, and immiscible. Organic solvents such as chloroform, hexane, 2-propanol, ethanol, and others are frequently employed. This approach is not commonly used to isolate proteins.

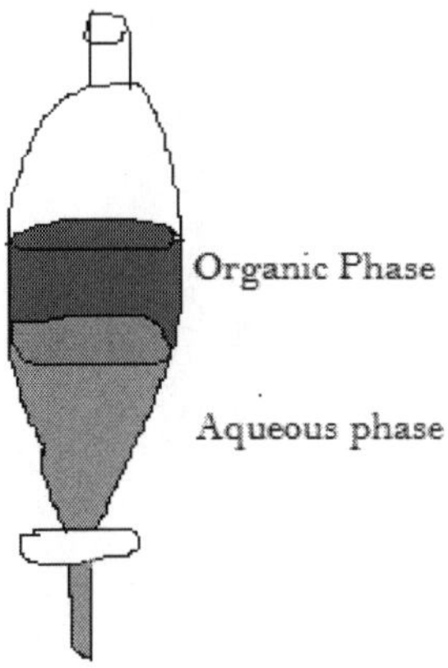

Figure 6.13. - Solvent extractor

In most cases, the broth produces 20–35 mg/l of penicillin. The clarified broth is typically treated with organic solvents such as amyl acetate or butyl acetate after the initial processing. The antibiotic is extremely soluble in solvents such as butyl acetate, amyl acetate, methyl isobutyl ketone, and methyl ethyl ketone, and can be utilized with any of them. Finally, penicillin is isolated as its sodium derivative by precipitation in a butanol–water mixture.

Recovery and purification of primary and secondary metabolites like acids and antibiotics with organic solvents, alcohols, carrier-modifier-solvent systems for downstream processing will be conducted in solvent extractor (Fig. 6.13). Solvent extraction is a common procedure since it is both effective and inexpensive (Shi et al., 2005). The solvents will be largely reused after the product has precipitated and crystallized.

Adsorption

Adsorption is a surface phenomenon in which the product adheres to an interface known as the adsorption system. The most common adsorption technique is ion-exchange adsorption. Many organic compounds are purified using this technique, which employs charcoal, porous polymers, and other adsorbents. The yield is proportional to the surface area and continues until saturation is attained. This is referred to as the breakthrough point. For any type or size of bed, standardization of breakthrough points is necessary. This will enable the protocols to be designed with clear indicators of the amount of adsorbent to be used, the time required for the procedure, and so on. Activated carbon is used to purify citric acid.

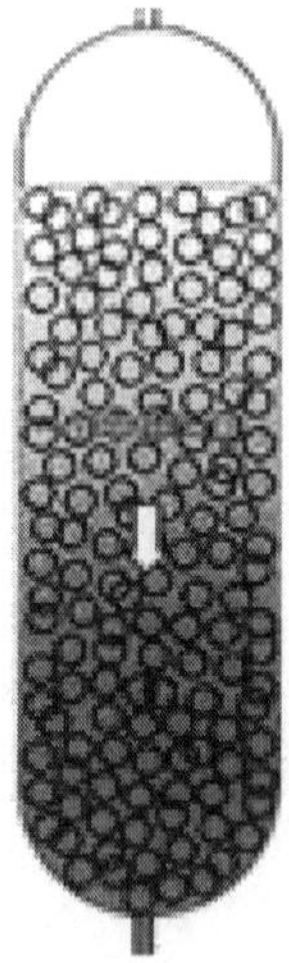

Figure 6.14. - Adsorption process

In the adsorption process, a bed of adsorbent column can be made and the culture broth can be passed through adsorbent column. Binding of the molecules to the surface is weak and reversible but when a molecule contains chromogenic group, they are strongly adsorbed on the activated carbon. Silica gel, activated carbon and alumina are commonly used as adsorbents in the downstream process during biosurfactant recovery (Fig. 6.14).

Product Precipitation

Product precipitation method is used to separate products that are in solution or have been dissolved in a solvent. Precipitation is accomplished through the addition of a chemical with a higher solubility and affinity to the solvent. Ammonium sulphate is widely used to precipitate proteins from their solutions because it is highly soluble, less toxic, and less expensive. The inorganic salt is gradually added to the protein solution until it approaches supersaturation, at which point the protein is eluted. The process of salting out is a sort of precipitation. Antibiotics are precipitated using butyl acetate, isobutyl ketone and mineral ions.

Ultrafiltration

Ultrafiltration approach is a powerful tool in molecular level process that uses several filters to separate particles with sizes ranging from 0.001 to 0.01m, such as proteins (Lobb et al., 2015). Ultrafiltration is basically a pressure-driven separation process. Multiple ultrafiltration units with varying filter capacity can facilitate in product isolation. Ultrafiltration aids in the removal of cell debris, bacteria, and viruses from broth, as well as the purification and concentration of molecules of interest. The filters are often flat or hollow and constructed of steel or ceramic. In the filtration units, these filter membranes are supported by a thick, very porous framework. Water and other solvents can be purified using this procedure, which is known as diafiltration (Fig. 6.15).

Several processes like buffer exchange and desalting, protein concentration, virus clearance, protein purification, and clarification in downstream processing is conducted by Ultrafiltration. The pore size between 10 and 500 Å via microfiltration and reverse osmosis as intermediate response to a given pressure driving force in this process. Membrane selectivity is mainly depending on hydrodynamic (e.g., transport) interactions and thermodynamic (e.g., partitioning).

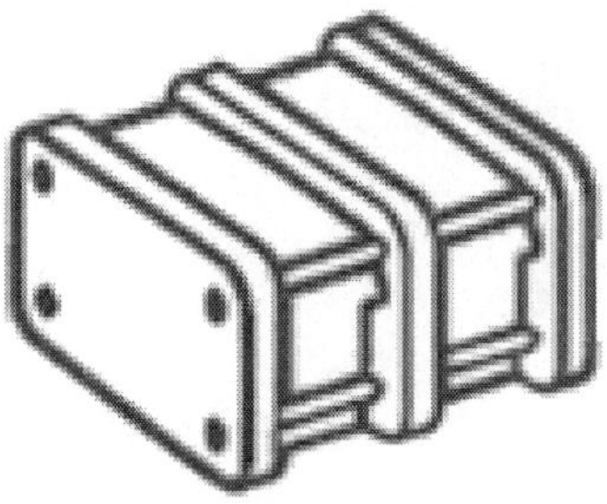

Figure 6.15. - Ultrafiltration

6.4 Product Purification

After eliminating the suspended particles and concentrating the product using the methods described above, the product must be purified further (Chase, 1994). This stage of downstream processing removes all impurities with properties similar to the products. The approaches discussed in this section are extensively used in the procedure. Chromatographic techniques are extremely successful and dominant in their ability to resolve mixtures and isolate the constituents. To purify high-value products such as vitamins, alkaloids, steroids, nucleic acids, amino acids, proteins, and so on, several chromatographic procedures are used.

The chromatography technique is based on the differential migration and selective retardation of solute molecules as they pass through the stationary phase (Hearn, 2017). The stationary phases differ depending on the type of chromatography and the molecule to be eluted. The mobile phase that transports the solutes across the column aids in elution. The mobile phases are chosen based on the eluted product and the stationary phase. The chromatographic methods are chosen based on the unique properties of the purified product or the remaining components to be removed.

Size Exclusion Chromatography

Size exclusion chromatography method separates molecules depending on their size, shape, and molecular weight (Pelekani et al., 1999). The uniformly porous gel beads function as molecular sieves, separating smaller and larger molecules. This approach is used to purify concentrated protein mixtures. A solution mixture containing molecules of varying sizes (e.g., different proteins) is added to the column and eluted.

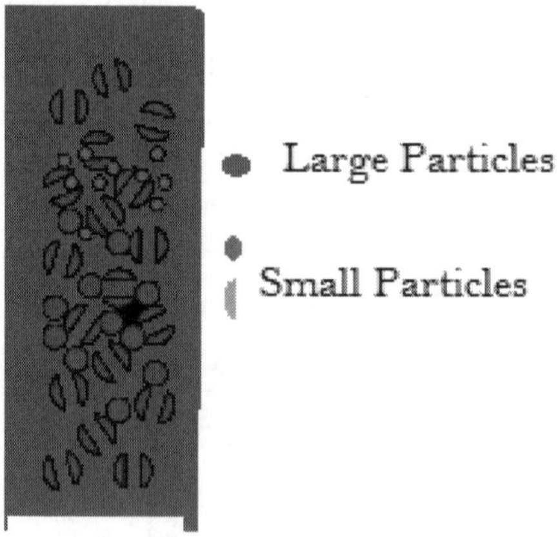

Figure 6.16. - Size exclusion chromatography

This is especially helpful for desalting protein concoctions. The stationary phase materials have set pore sizes and are typically comprised of cellulose, agarose, acrylic polymers, and other compounds. The choice of matrix material is mostly dependent on their reactivity to the target product. Solute molecules in the pore size range move in and out of the beads. Molecules larger than the pore size pass through the interstitial spaces and flow down the column faster. Later on, the smaller molecules in decreasing order of size are eluted. The low molecular weight compounds and salts are efficiently removed from the purified product through this method (Fig. 6.16).

Previous studies were shown that size exclusion chromatography methods in downstream purification method of extracellular vesicles (EVs) for helminth pathogens are good in compared to a differential centrifugation method (Davis et al., 2019).

Ion-exchange chromatography

Ion-exchange chromatography method separates molecules based on their surface charges (Fekete et al., 2015). It is effective against any charged molecule, including proteins, nucleic acids, and amino acids. The pH of the eluting molecules, as well as the matrix, plays an important role in the process since it dictates their effective charge. This approach allows for two types of ion exchange separation depending on the matrix material employed. In anion exchange chromatography, the stationary phase is positively charged, whereas in cation exchange chromatography, it is negatively charged. As a result, the stationary phase is determined by the charge of the target molecule to be separated. Dowex HCR and Amberlite IR are the most regularly used cation exchangers, whereas Dowex SAR and Amberlite IRA are the most commonly used anion exchangers. By varying the pH of the eluant or increasing the concentration of the salt solution, the ionic bound molecules can be eluted from the matrix. Proteins with few charged groups are eluted at low salt concentrations, while proteins with many charged groups are eluted at high salt concentrations. Antibiotics and proteins can be purified using ion-exchange chromatography (Fig. 6.17).

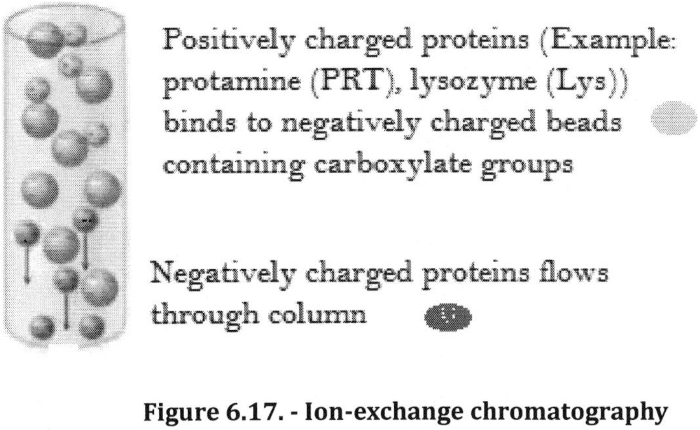

Figure 6.17. - Ion-exchange chromatography

Affinity chromatography

Affinity chromatography method isolates molecules based on their specific binding interactions with the immobilized ligand in the column (Hage, 1999). Affinity chromatography is a prominent downstream processing approach for purifying various types of biopharmaceuticals due to its high selectivity and resolution over other chromatographic techniques (Fig. 6.18; Table 6.2). This approach is commonly used for protein purification. This technique is used to separate both the natural proteins and specifically designed recombinant proteins with tag sequences. The ligands used are determined by the molecule of interest that needs to be purified, such as enzyme and substrate; enzyme and inhibitor; antibody and antigen; ligand and receptor, and so on. The popularly used affinity tags are histidine and Glutathione – s-transferase.

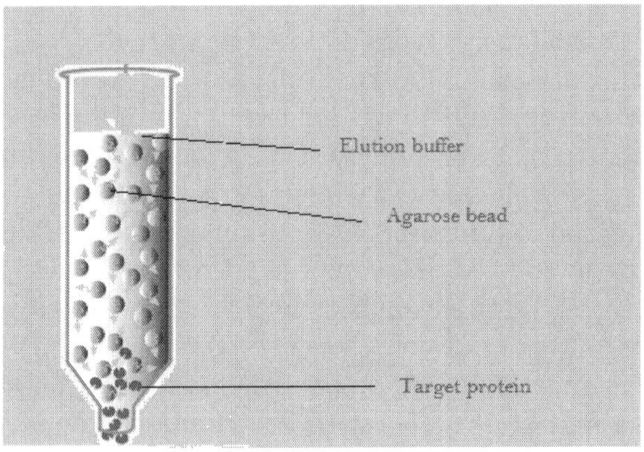

Figure 6.18. - Affinity chromatography

Table 6.2. - Ligands used for separation of proteins in Affinity chromatography

Ligand	Protein
Antibody	Antigen
Receptor	Hormone
Hepten	Antibody
Cofactor	Enzymes
Inhibitor	Enzyme
Lectins	Glycoproteins
Metal ions	Metal ion binding proteins
Heparin	Coagulating factors

The histidine tag-containing protein and other contaminants are run through a specifically designed affinity column that immobilizes Nickel or any transition metal ion (Kabir et al., 2010). The histidine tag binds to the metal ions, and the remainder of the mixture is drained. The attached protein is eluted selectively with imidazole, which has a greater affinity for metal ions. The chromatographic techniques and principle of separation were presented in Table 6.3.

Table 6.3. - Chromatographic techniques and principle for separation of proteins

Chromatography	Principle
Size exclusion chromatography	Size, shape, and molecular weight
Ion-exchange chromatography	Net charge
Affinity chromatography	Biological affinity and molecular recognition

Fractional Precipitation

Fractional precipitation technique separates the molecules from the solution based on their difference in solubilities (Cortez and Pessoa, 1999). The compounds having a common ion/s but different solubility constants can be separated by fractional precipitation. Any compound, especially protein, will precipitate across a range of solution ionic strengths, and different proteins will precipitate over different ranges. Using this technique, it is possible to purify a protein at times but more commonly will concentrate the protein of interest. It is extremely effective for isolating a collection of proteins. The phosphoproteins are selectively precipitated by trivalent europium, terbium, and erbium metal ions from the milk or egg. Equally, the alcohol dehydrogenase is fractionally precipitated from clarified baker's yeast homogenate using saturated ammonium sulphate solution.

Anions like Cl−, Br−, CO2−3, SO2−4, Cr2O2−4, S2−, PO2−4, OH− etc. are used for the separation of mixture of metal ions in a solution by precipitation. Based on heterogeneous **equilibrium** phenomenon, the solubilities and specific precipitation reactions of two metal salts vary and can be separated by fractional precipitation (Fig. 6.19).

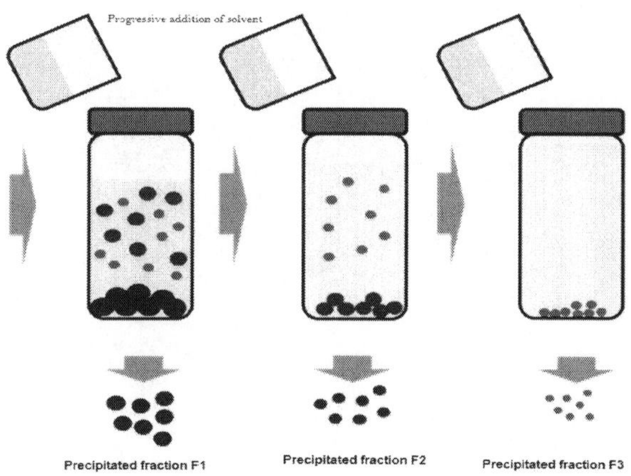

Figure 6.19. - Fractional Precipitation

6.5 Product Polishing or Finishing

Product polishing or finishing is the final stage of the process in which the purified compound's stability, transportability, and other properties are improved (Khan, 2018). The excess moisture is eliminated without altering the compound's activity or stability. Most compound solutions are concentrated by removing moisture content, however certain small molecule solutions are concentrated through crystallization. Moreover, the highly sensitive compounds are stabilized by the addition of various stabilizers such as sugars, salts, polyhydric alcohols, and so on.

Lyophilization or Freeze Drying

This technique concentrates heat labile chemicals. The principle of lyophilization is sublimation of a liquid from the frozen state and then dried under vacuum in a freeze-dryer. It works by sublimating the liquid from the frozen condition, thereby eliminating the solvents. The solution containing the molecule of interest is frozen and then dried in a vacuum during the process (Fig. 6.20). The low temperature used in the procedure has no effect on the biological activity of the products such as antibiotics, proteins, enzymes, and so on. This procedure also concentrates low molecular weight compounds that are too small for filtration. This is the most commonly used process for drying and formulating items in the pharmaceutical, diagnostic, and food industries.

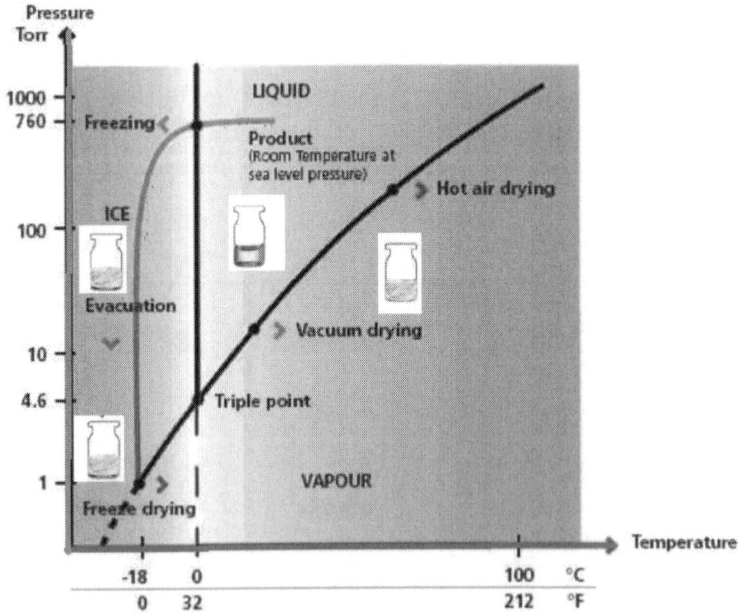

Figure 6.20. - Freeze drying

Spray drying

In this procedure, heat stable chemicals are concentrated and made suitable for handling and storage. It operates by vaporizing the solvent at high temperatures. The solution containing the product of interest is introduced into the heat chamber as small droplets. When the mist droplets come into contact with the hot airflow inside the chamber, the moisture evaporates. As the solvent evaporates, the product remains solid and is recovered from the cyclone. There are many advantages of this method. 1) The product solution need not be concentrated. 2) The drying time is extremely low. 3) The product comes in small grain size. Spray drying is commonly used in the food and pharmaceutical industry (Fig. 6.21).

Figure 6.21. - Spray drying

Crystallization

Crystallization method allows the product to convert from a solution to a solid crystalline state while avoiding the amorphous form (Fig. 6.22). Temperature modulation generates the product crystals; however, the most prevalent mechanism of crystallization is product precipitation. This is accomplished by supersaturating the solution with salts, which decreases the solubility of the product in the solvent. Antibiotics and organic acids are concentrated by this process.

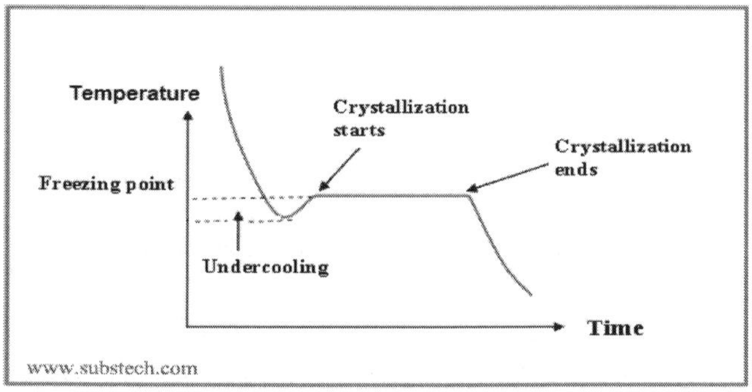

Figure 6.22. - Crystallization zone

6.6 Importance in Industrial Market

In the present decades, biotechnology industry employs mammalian cells, plant cells, recombinant bacteria, and transgenic animals for the production of several high-value therapeutic proteins like insulin, interferon's, monoclonal antibodies, etc (Kalyanpur, 2002). Proteins constitute a large class of amphoteric biopolymers with molecular masses ranging from 5 to 20000 kDa, which are based on amino acids as building blocks. Proteins are well tolerated, can be highly potent, and often posses a long half - life after administration, making them effective therapeutics. The biopharmaceutical market share in the industrial productions is shown in figure 6.23.

Downstream processing is used to isolate, purify and concentrate the desired products from the complex system during processing of from drug substance to drug product. It is very important in pharmaceutical industries because the entire bioprocess is influenced by many factors that is associated recovery costs. The equipment also covers a huge range of complex working systems varying in size and complexity. The recovery and purification of industrial products are highly an expensive process in pharmaceutical industries and are most complicated and complex process in the industry.

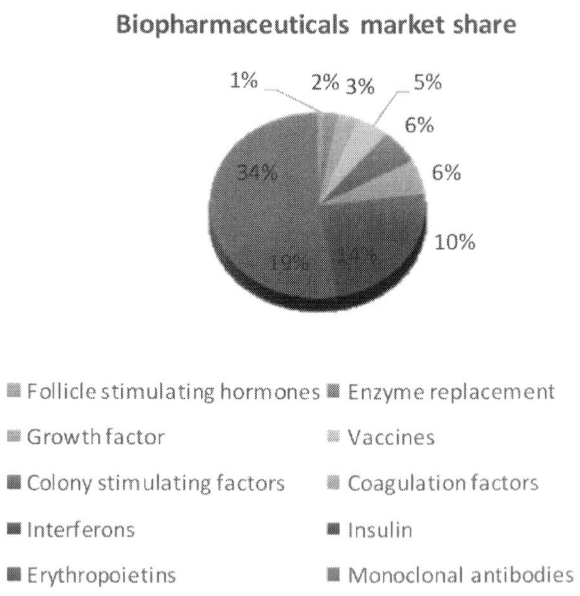

Figure 6.23. - Biopharmaceutical market share

Table 6.4 shows the industrial methods are used in protein therapeutics for quality assurance. Hence, continuous downstream processing in industries showing high value biological products is very much essential for processing of therapeutic proteins (Yilmaz et al., 2020).

Table 6.4. - Protein therapeutics for quality assurance

Analytical technique	*Impurity / Contaminant*
SDS-PAGE Electrophoresis	Protein Contaminant, Proteolytic degradation products
HPLC	Protein Contaminant like Host cell proteins, Proteolytic degradation products
DNA Dot blot hybridization	DNA
Sterility test	Microbial contamination
Rabbit pyrogen test	Endotoxin

Continuous manufacturing of industrial product has been applied in several industries but has been implemented in production of biotechnology product that are helpful for human community (Jungbauer, 2013). Filtration and chromatography are the central unit operation that has been adopted and used in many continuous operation systems. The driving forces for the implementation of continuous manufacturing strategies in the present decades are parametric release considerations, economic pressure and flexibility in the industries. Flexible and generic downstream process platform for monoclonal antibodies and Fc fusion proteins (mAbs) through Protein A chromatography is a highly selective step in downstream process (Shukla et al., 2007).

The quality criteria that are obligatory in several biochemicals are stringent with high-separation and purification technologies like Affinity-based purification that is important to downstream processing (Labrou and Clonis, 1994). The production of microalgal biofuels, monoclonal antibodies (mAbs), microbial alkaline proteases and proteins, polyunsaturated fatty acids from microalgae are some of the important industrial products that are having huge demands in the industrial markets. Hence downstream processing is a bottleneck process in the manufacturing scheme that may also use genetic or biochemical engineering for the development of novel products in industries with application of different types of microorganisms (Somasundaram et al., 2018; Bhatia and Paliwal, 2010). The industrial microbiological techniques that were discussed in the present chapter provides better understanding on downstream processing that is applicable in the biological industries.

6.7. Conclusion

Downstream processing is an important process in production of industrial products in the present decades. There are several healthcare and commercial products that are important in daily needs of humans. From the process of filtration to the advanced methods like HPLC and electrophoresis, lot of large scale and small-scale productions are conducting in the production industries related to healthcare systems.

Acknowledgements

The authors like to thank administration of Atal Bihari Vajpayee University, Bilaspur (CG) and IGNTU, Amarkantak for providing assistance and support.

References

Anastassiadis, S. (2007). L-Lysine fermentation. Recent patents on Biotechnology, 1(1), 11-24.

Araùjo, M., Hube, L. A. & Stasyk, T. (2008). Isolation of endocitic organelles by density gradient centrifugation. 2D PAGE: Sample Preparation and Fractionation, 317-331.

Bajpai, P. (2014). Xylanolytic enzymes. Academic press.

Bals, B., Rogers, C., Jin, M., Balan, V. & Dale, B. (2010). Evaluation of ammonia fibre expansion (AFEX) pretreatment for enzymatic hydrolysis of switchgrass harvested in different seasons and locations. Biotechnology for biofuels, 3(1), 1-11.

Bhatia, L. & Paliwal, S. (2010). Banana peel waste as substrate for ethanol production. International Journal of Biotechnology and Bioengineering Research, 1(2), 213–218.

Bhatia, L., Singh, A., Chandel, A. & Singh, O. M. (2018). Biotechnological Advancements in cellulosic ethanol production. O. V. Singh and A. K. Chandel (eds.), Sustainable Biotechnology- Enzymatic Resources of Renewable Energy, https://doi.org/10.1007/978-3-319-95480-6.

Bergemann, K., Christian, E. and Sandra P. (2007). Production and downstream processing. Handbook of therapeutic antibodies, 1, 199-238.

Bowen, W. R. & Jenner, F. (1995). Theoretical descriptions of membrane filtration of colloids and fine particles: an assessment and review. Advances in Colloid and Interface Science, 56, 141-200.

Branyikova, I., Prochazkova, G., Potocar, T., Jezkova, Z. & Branyik, T. (2018). Harvesting of microalgae by flocculation. Fermentation, 4(4), 93.

Brauchle, E. & Schenke-Layland, K. (2013). Raman spectroscopy in biomedicine–non-invasive in vitro analysis of cells and extracellular matrix components in tissues. Biotechnology journal, 8(3), 288-297.

Brinkmann, A., Elouafiq, S., Pieracci, J. & Westoby, M. (2018). Leveraging single-pass tangential flow filtration to enable decoupling of upstream and downstream monoclonal antibody processing. Biotechnology progress, 34(2), 405-411.

Burden, D. W. (2012). Guide to the disruption of biological samples-2012. Random Primers, 12(1), 1-25.

Caulet, P. J. C., Van der Lans, R. G. J. M. & Luyben, K. C. A. (1996). Hydrodynamical interactions between particles and liquid flows in biochemical applications. The Chemical Engineering Journal and The Biochemical Engineering Journal, 62(3), 193-206.

Chase, H. A. (1994). Purification of proteins by adsorption chromatography in expanded beds. Trends in biotechnology, 12(8), 296-303.

Chisti, Y. (1998). Strategies in downstream processing. Bioseparation and bioprocessing: a handbook, 2, 3-30.

Cortez, E. V. & Pessoa Jr, A. (1999). Xylanase and β-xylosidase separation by fractional precipitation. Process Biochemistry, 35(3-4), 277-283.

Dai, J. Y., Sun, Y. Q. & Xiu, Z. L. (2014). Separation of bio-based chemicals from fermentation broths by salting-out extraction. Engineering in Life Sciences, 14(2), 108-117

Davis, C. N., Phillips, H., Tomes, J. J., Swain, M. T., Wilkinson, T. J., Brophy, P. M. & Morphew, R. M. (2019). The importance of extracellular vesicle purification for downstream analysis: A comparison of differential centrifugation and size exclusion chromatography for helminth pathogens. PLoS neglected tropical diseases, 13(2), e0007191.

Didaras, N. A., Karatasou, K., Dimitriou, T. G., Amoutzias, G. D. & Mossialos, D. (2020). Antimicrobial activity of bee-collected pollen and beebread: State of the art and future perspectives. Antibiotics, 9(11), 811.

Diels, A. M. & Michiels C. W. (2006). High-pressure homogenization as a non-thermal technique for the inactivation of microorganisms. Critical reviews in microbiology, 32(4), 201-216.

Ditu, L. M. & Gheorghe, I. (2017). Introduction in soft chemistry and food fermentation. In Soft Chemistry and Food Fermentation (pp. 1-19). Academic Press.

Fekete, S., Beck, A., Veuthey, J. L. & Guillarme, D. (2015). Ion-exchange chromatography for the characterization of biopharmaceuticals. Journal of pharmaceutical and biomedical analysis, 113, 43-55.

Galanakis, C. M. (2013). Emerging technologies for the production of nutraceuticals from agricultural by-products: a viewpoint of opportunities and challenges. Food and Bioproducts Processing, 91(4), 575-579.

Grima, E. M., Fernández, F. A. & Medina, A. R. (2004). 10 Downstream Processing of Cell-mass and Products. Handbook of microalgal culture: biotechnology and applied phycology, 215.

Gupta, R., Beg, Q., Khan, S. & Chauhan, B. (2002). An overview on fermentation, downstream processing and properties of microbial alkaline proteases. Applied microbiology and biotechnology, 60(4), 381-395.

Hage, D. S. (1999). Affinity chromatography: a review of clinical applications. Clinical chemistry, 45(5), 593-615.

Hammerschmidt, N., Hobiger, S. & Jungbauer, A. (2016). Continuous polyethylene glycol precipitation of recombinant antibodies: sequential precipitation and resolubilization. Process Biochemistry, 51(2), 325-332.

Hao, J., Xu, F., Liu, H. & Liu, D. (2006). Downstream processing of 1, 3-propanediol fermentation broth. Journal of Chemical Technology & Biotechnology:

International Research in Process, Environmental & Clean Technology, 81(1), 102-108.

Hearn, M. T. (2017). High-performance liquid chromatography of peptides and proteins: General principles and basic theory. In High-Performance Liquid Chromatography of Peptides and Proteins: Separation, Analysis, and Conformation (pp. 95-104). CRC Press.

Hwang, K. J., Chan, C. S. & Tung, K. L. (2009). Effect of backwash on the performance of submerged membrane filtration. Journal of membrane science, 330(1-2), 349-356.

Jungbauer, A. (2013). Continuous downstream processing of biopharmaceuticals. Trends in biotechnology, 31(8), 479-492.

Kabir, M.E., Senthilkumar, K., Masahiko, M., Yasuhiro, F., & Tadazumi, K. (2010). Purification and functional characterization of a Camelid-like single-domain antimycotic antibody by engineering in affinity tag. Protein expression and purification. 72(1), 59-65.

Kaladhar, DSVGK. (2014). Industrial Biotechnology. Sonali Publications, India

Kalyanpur, M. (2002). Downstream processing in the biotechnology industry. Molecular biotechnology, 22(1), 87-98.

Kazemi, A. S. & Latulippe, D. R. (2014). Stirred well filtration (SWF)–A high-throughput technique for downstream bio-processing. Journal of membrane science, 470, 30-39.

Khan, F. A. (2018). Industrial biotechnology. In Biotechnology Fundamentals (pp. 493-530). CRC Press.

Khanra, S., Mondal, M., Halder, G., Tiwari, O. N., Gayen, K. & Bhowmick, T. K. (2018). Downstream processing of microalgae for pigments, protein and carbohydrate in industrial application: A review. Food and bioproducts processing, 110, 60-84.

Kleerebezem, R. & van Loosdrecht, M. C. (2007). Mixed culture biotechnology for bioenergy production. Current opinion in biotechnology, 18(3), 207-212.

Kliaugaitė, D., Yasadi, K., Euverink, G. J., Bijmans, M. F. & Racys, V. (2013). Electrochemical removal and recovery of humic-like substances from wastewater. Separation and Purification Technology, 108, 37-44.

Labrou, N. & Clonis, Y. D. (1994). The affinity technology in downstream processing. Journal of biotechnology, 36(2), 95-119.

Lee, S. Y., Chew, K. W. & Show, P. L. (2019). Cell separation and disruption, product recovery, and purification. In Essentials in fermentation technology (pp. 237-271). Springer, Cham.

Leung, W. W. F. (2020). Centrifugal separations in biotechnology. Butterworth-Heinemann

Lindner, J. (2014). On Continuous Magnetically Enhanced Centrifugation in Large Scale Downstream Processing of Bioproducts (Doctoral dissertation, Universität Karlsruhe).

Lobb, R. J., Becker, M., Wen Wen, S., Wong, C. S., Wiegmans, A. P., Leimgruber, A. & Möller, A. (2015). Optimized exosome isolation protocol for cell culture supernatant and human plasma. Journal of extracellular vesicles, 4(1), 27031.

Lohmann, L. J. & Strube, J. (2020). Accelerating biologics manufacturing by modeling: process integration of precipitation in mAb downstream processing. Processes, 8(1), 58.

Löwer, E., Pham, T. H., Leißner, T. & Peuker, U. A. (2020). Study on the influence of solids volume fraction on filter cake structures using micro tomography. Powder Technology, 363, 286-299.

Lute, S., Kozaili, J., Johnson, S., Kobayashi, K. & Strauss, D. (2020). Development of small-scale models to understand the impact of continuous downstream bioprocessing on integrated virus filtration. Biotechnology progress, 36(3), e2962.

Ma, F. & Hanna, M. A. (1999). Biodiesel production: a review. Bioresource technology, 70(1), 1-15.

Manikandan, N. A., Pakshirajan, K. & Pugazhenthi, G. (2019). A novel ceramic membrane assembly for the separation of polyhydroxybutyrate (PHB) rich Ralstonia eutropha biomass from culture broth. Process Safety and Environmental Protection, 126, 106-118.

Medina, A. R., Grima, E. M., Giménez, A. G. & González, M. I. (1998). Downstream processing of algal polyunsaturated fatty acids. Biotechnology advances, 16(3), 517-580

Musa, M., Ward, A., Ayoko, G. A., Rösch, C., Brown, R. & Rainey, T. J. (2020). Single-step dynamic dewatering of microalgae from dilute suspensions using flocculant assisted filtration. Microbial cell factories, 19(1), 1-18.

Nitsos, C., Filali, R., Taidi, B. & Lemaire, J. (2020). Current and novel approaches to downstream processing of microalgae: A review. Biotechnology Advances, 107650.

Pelekani, C., Newcombe, G., Snoeyink, V. L., Hepplewhite, C., Assemi, S. & Beckett, R. (1999). Characterization of natural organic matter using high performance size exclusion chromatography. Environmental Science & Technology, 33(16), 2807-2813.

Saboyainsta, L. V. & Maubois, J. L. (2000). Current developments of microfiltration technology in the dairy industry. Le Lait, 80(6), 541-553.

Sankaran, S., Khanal, S. K., Jasti, N., Jin, B., Pometto III, A. L. & Van Leeuwen, J. H. (2010). Use of filamentous fungi for wastewater treatment and production of high value fungal byproducts: a review. Critical reviews in environmental science and technology, 40(5), 400-449.

Scholz, M. (2015). Wetland systems to control urban runoff. Elsevier.

Shi, J., Nawaz, H., Pohorly, J., Mittal, G., Kakuda, Y. & Jiang, Y. (2005). Extraction of polyphenolics from plant material for functional foods—Engineering and technology. Food reviews international, 21(1), 139-166.

Shukla, A. A., Hubbard, B., Tressel, T., Guhan, S. & Low, D. (2007). Downstream processing of monoclonal antibodies—application of platform approaches. Journal of Chromatography B, 848(1), 28-39.

Somasundaram, B., Pleitt, K., Shave, E., Baker, K. & Lua, L. H. (2018). Progression of continuous downstream processing of monoclonal antibodies: Current trends and challenges. Biotechnology and bioengineering, 115(12), 2893-2907.

Taniguchi, M., Kotani, N. & Kobayashi, T. (1987). High-concentration cultivation of lactic acid bacteria in fermentor with cross-flow filtration. Journal of Fermentation Technology, 65(2), 179-184.

Tsafrakidou, P., Michaelidou, A. M. & G Biliaderis, C. (2020). Fermented cereal-based products: Nutritional aspects, possible impact on gut microbiota and health implications. Foods, 9(6), 734.

Wang, Y., Ling, C., Chen, Y., Jiang, X. & Chen, G. Q. (2019). Microbial engineering for easy downstream processing. Biotechnology advances, 37(6), 107365.

Watson, J. H. (1973). Magnetic filtration. Journal of Applied Physics, 44(9), 4209-4213.

Yesil-Celiktas, O. & Vardar-Sukan, F. (2013). Downstream processes for plant cell and tissue culture. In Biotechnology for Medicinal Plants (pp. 1-27). Springer, Berlin, Heidelberg

Yilmaz, D., Mehdizadeh, H., Navarro, D., Shehzad, A., O'Connor, M. & McCormick, P. (2020). Application of Raman spectroscopy in monoclonal antibody producing continuous systems for downstream process intensification. Biotechnology progress, 36(3), e2947.

Yokoi, H., Natsuda, O., Hirose, J., Hayashi, S. & Takasaki, Y. (1995). Characteristics of a biopolymer flocculant produced by Bacillus sp. PY-90. Journal of Fermentation and Bioengineering, 79(4), 378-380.

Chapter-7

FRONTIERS OF MICROORGANISMS IN INDUSTRIAL PROCESS

Akanksha Kulshreshtha[1], *Ajay Kumar Kataria*[1], *Soumya Sasmal*[1] *and O.N. Tiwari*[2]*

[1] *Department of Biological Sciences and Engineering, Netaji Subhas University of Technology, New Delhi.*

[2]*Centre for Conservation and Utilisation of Blue Green Algae (CCUBGA), Division of Microbiology, ICAR-Indian Agricultural Research Institute (IARI), New Delhi.*

**Corresponding author Email: ontiwari1968@gmail.com*

Abstract

Microorganisms are essential to life on Earth. For nearly a century, microbes have assisted in industrial research. They have given us enzymes and metabolic pathways with greater variety. Genetic engineering has altered industrial microbiology greatly. A wide range of techniques has been used to create novel metabolic pathways that produce new metabolites, modify the biosynthesis pathways to supply biosynthetic intermediates, and give enzymes with enhanced selectivity, higher catalytic activity, and greater stability. For centuries, it has been a secret weapon, protecting milk and produce while providing ingredients for the foods we eat every day, including bread, beer, wine, distilled spirits, vinegar, cheese, and pickles. Industrial microbiology underwent a second important phase in the early 20th century. The first large-scale commercial fermentations were devoted to the synthesis of solvents, organic acids, vitamins, enzymes, and other goods during this golden age of fermentation. In the middle of the 20th century, fermentation techniques containing antibiotics such as penicillin and streptomycin were developed. A new branch of biochemistry (microbiology) engineering emerged from studies on the scale-up of these two antibiotics. This new strain enhancement technology, which was based on scholarly research into mutational microbial genetics, followed directly on the heels of the discovery. It is increasingly common for the genomes of industrial microbes to be sequenced, and this is providing extensive insight into the genetic and enzymatic makeup of these invaluable species. A variety of important technologies such as functional genomics, proteomics, and metabolomics are currently being used to develop new medication compounds and catalyst enzymes.

Keywords: *Microorganisms, fermentation, enzyme, industry, culture, metabolic engineering.*

7.1. Introduction

Microbes maintain life on earth through their many connections and biogeochemical processes. However, they are not necessarily irreproducible in their roles. If someone says that we can survive without micro-organisms, one should request that they should qualify for this statement. Will life be extinguished if there are no microorganisms or if there are no germs? Not all life will be lost immediately or for a long time. What if the antimicrobial wand was waved once more, this time wiping out all microbes (algae, archaea, bacteria, fungi, viruses, etc.) from the planet? The startling absence of all kinds of microbial diseases, such as Ebola, ulcers, athlete's foot, to name a few, would be one of the first impressions in a world free of all microbes. If bacteria and archaea were removed from the earth, the future of life on our planet would not be much like our current world without bacteria and archaea. It would be far more drastic impact on the environment. In the absence of microbes' human and animal waste would accumulate quickly, and other bio-geochemical cycles would cease.

Even fungi and protists are incapable of the task, microorganisms are the only organisms that are capable of recycling biomass in terrestrial and aquatic environments, causing rapid depletion of available macronutrients and micronutrients (Demain and Adrio, 2008). Living food sources would become increasingly scarce. Even without the presence of microorganisms it is not possible for us to digest food.

Without microbial symbionts, most ruminant cattle would hunger, and plants would rapidly deplete nitrogen, cease photosynthesis, and die. In the face of rising environmental, ecological, and humanitarian crisis, the intensive human intervention required to generate and distribute sufficient vitamins, plant fertilizers, and food sources will likely overtake inventiveness. Small pockets of people and other creatures (e.g., insects) would persist for a time, perhaps decades or centuries, but the long-term survival of most eukaryotes would be dubious (Gilbert and Neufeld., 2014).

Therefore, microbes are necessary for life on earth because they perform a variety of critical and useful biological tasks. Photosynthesis is the process through which algae and bacteria absorb energy from the sun and convert it to food, establishing the foundation of the food chain. Many microbes decompose dead and decayed substances, recycling nutrients that can be utilized by other organisms.

Nitrogen Fixation: Some bacteria are capable of absorbing nitrogen from the air and incorporating it into the soil.

7.2. World of Microorganism

Humans exist during a teeming, living microbiome. Microorganisms can be found in every imaginable habitat like ranging from the surface of human skin to rainforest soils to hydrothermal vents on the ocean floor. Despite their little size, microorganisms nonetheless have a huge impact on our environment. Bacteria are an essential component of the biosphere because of their impact on both evolutionary and ecological processes. The discovery of microorganisms' important role in biogeochemical cycling, pollution removal, wastewater treatment, and oxygen generation has shed considerable light on the role of microorganisms in all ecosystem functions (Hayat et al., 2010; Senni et al., 2011).

Microbes and microbial products are so significant for bioenergy and biofuel production. They are a natural source of new medicines and are utilized as biopesticides and biofertilizers, as well as to help preserve the environment's sustainability (Hayat et al., 2010). Despite their beneficial role in the environment, microorganisms also function as pathogens and have an impact on human health and hygiene (Arrigo, 2005; Lal et al., 2010; Singh et al., 2010; Kostka et al., 2011).

Microbial diversity is quite remarkable. Some microbes thrive in scorching temperatures, whereas others can live in frozen seas. Some types of microbes can generate nitric and sulfuric acids. Anaerobic microbes carry out many vital functions in the ecosystem that plants and animals cannot, including methane generation and nitrogen fixation. Key processes in bread making, yoghurt manufacturing, making pickles, wine, beer, and cheese rely on microbes (Staley et al., 1997). The foundation of biotechnology is microbiology. Antibiotics and anti-cancer medicines include penicillin, streptomycin, and chloramphenicol are produced from microbes. A greater need for antibiotic medication has emerged because of the growth of dangerous pathogens that are becoming resistant to treatment. Microorganisms have various species, each of which is a source of several kinds of goods, including biotechnology applications and potential uses that are currently unknown (Staley et al., 1997).

The tremendous capacity of microorganisms to adapt and resist to environmental disruption is one of the main features of the microorganisms, especially the prokaryotes (Allison and Martiny, 2008). This is done by alteration of their physiology and genome transformation by mutation and gene acquisition by horizontal gene transfer (Sitaraman, 2018).

7.3. Preservation and Culture Collection of Microbes

Bacterial cultures are heavily skewed toward a few phylogenetic groupings. Many of the currently unknown bacterial species are anticipated to contain novel biosynthetic pathways and unknown metabolic characteristics. Based on a better understanding of the ecology of hitherto uncultured bacteria, new cultivation concepts have been devised. Improved mediums that mirror the natural types and quantities of substrates and nutrients, high-throughput cultivation techniques, and tactics that take advantage of biofilm development and bacterial interactions have all been particularly successful (Zhang et al., 2021). Metagenomics and single-cell genomics can disclose previously undiscovered metabolic properties of uncultured bacteria and, when combined with culture-independent physiological assessments, can aid in more efficient targeting of functional innovation (Gutleben et al., 2018). However, many novel species of bacteria that were first enriched eluded separation. To recover and use a larger fraction of microbial diversity, future cultivation studies will need to focus on enhanced sub cultivation, purification, and preservation strategies.

Although it is well established that microbes play a critical role in ecosystem and biosphere health and diversity, Microorganisms have a high instrumental value, which refers to their beneficial effects on the biosphere and on people (Gupta et al., 2017). They contribute significantly to biogeochemical transformations in most habitats inhabited by multicellular organisms (Konopka, 2009). We rely on microbes, which is one reason to maintain and protect them. They are not just beneficial; they are necessary for our survival. From a more practical standpoint, microbial biological diversity also contains enormous quantities of valuable resources for the pharmaceutical, biotechnology, and food sectors (Bull, 2004; Challis, 2008). Microbial variety loss entails a loss of potentially important resources.

7.3.1. Preservation Techniques

In order to retain the organism alive, uncontaminated, and without variation or mutation, the major goal of culture preservation is to keep the culture in its original isolated form in terms of phenotype, genotype and growth conditions. Microorganisms must be preserved in a way such that, there is reduction in the metabolism of an organisms either by desiccation of the medium of growth, or by lowering the temperature of growth of an organism or by nutritional limitation for the growth (Tedeschi and De Paoli, 2011).

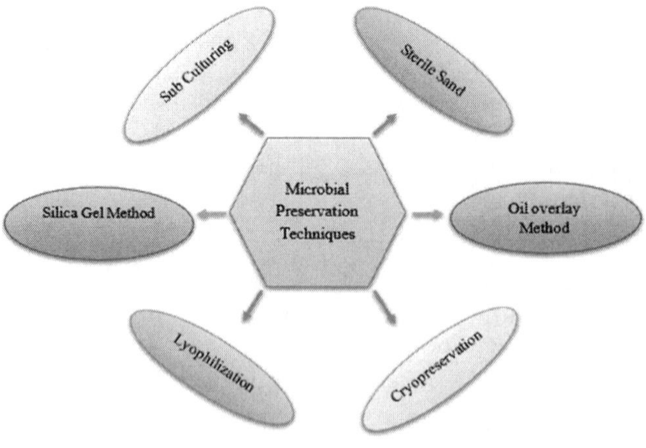

Figure 7.1 - Microbial Preservation Techniques

Several methods for preserving microorganisms have been successfully used for microbial preservation as shown in figure 7.1. Two most used long-term preservation techniques for microorganisms are Lyophilization and cryoprotection (Prakash et al., 2013). The decision between the two approaches is determined by in-house facilities, also by the features of the target organism, the purpose for preservation, and the period of storage.

7.3.2. Culture Collection

As pertaining to microbial resources, culture collections serve a critical role in conservation and their sustainable use (Daniel and Prasad, 2010). The collections also include reference strains, quality control tools, and other biological resources for high-quality research and education (Smith, 2003). Significance and Role played by microbial culture collection centers.

The breakthroughs in molecular biology have led to the further identification of novel microbial taxa and strains, which must be maintained to be available to researchers, teachers, and biotechnology workers. Due to lack of financial assistance and workforces, individual laboratories cannot do so. A cultural collection therefore plays very significant role as discussed in figure 7.2 (Sharma and Shouche, 2014).

Figure 7.2. - Role played by microbial culture collection centers

7.4. Potential Threats to Host Associated Microbial Diversity

Conservation, biologists have highlighted environmental contamination or pollution, infectious diseases, climate change etc. (Convention on Biological Diversity (CBD), 2010) as amongst the most important and prevalent biodiversity threats to our world. Contaminants in the environment can change host-associated microbial communities by displacing indigenous bacterial taxa with those that can tolerate continuous exposure to harmful substances (Costa et al., 2016).

When bacteria extracted from the skin of frogs living in acid mine drainage-contaminated environments were challenged with toxic effluent in the laboratory, they demonstrated considerably higher tolerance than germs derived from frogs residing at several uncontaminated reference locations. Contamination also affects the makeup of the microbial environmental communities (Fields et al., 2005), therefore affecting the reservoir of microorganisms present on the skin or in the gut of living organisms. Contamination of the environment can have a direct influence on host health by causing the loss of functionally essential microorganisms.

Climate change (increased temperature, acidity, and so on) can have a direct influence on microbial reservoirs accessible for host colonization by changing ambient microbial populations (Fields et al., 2005; Zogg et al., 1997). Furthermore, these influences may have an indirect effect on host-associated microbial populations due to their effects on host physiology. For example, when corals are stressed due to heat and acidification, they can produce antibacterial chemicals, which may affect the diversity or composition of their associated microbial communities. Ramsby et al. (2018) showed in sponges that when water temperature increased, key microbial taxa were lost, and once temperature reverted to baseline values, individuals failed to recover those symbionts (Ramsby et al., 2018). Infectious diseases might affect their host-associated microbial populations is through direct interactions between pathogens and symbionts. Changes in the microbial populations associated with the host may impair normal physiological function, ultimately affecting host health. Knutie, (2018) established that birds afflicted with parasite larvae have a decreased microbial diversity in their guts, which is associated with decreased antibody levels in these animals (Knutie, 2018).

7.4.1. Priorities for Conservation and Examples

Reason for protecting microorganisms are well established, at the same time conservation of microbes in their natural habitat is a complex process. Because of the importance biogeochemical cycles to the rest of the biosphere, microorganisms that play a role in global biogeochemical cycles are a high priority for conservation. These communities include ocean sediment communities and ocean phytoplankton, many of which are threatened by anthropogenic activities (Cockell and Jones, 2009). Microorganisms' critical involvement in global geochemical cycles demonstrates the need of microbial study in appraising purposeful intervention in these systems. Large-scale initiatives, such as iron fertilization of the seas, have been considered as a way of artificially increasing carbon sequestration (Boyd et al., 2007). These initiatives have the potential to have unexpected repercussions for microbial ecosystems. Assessing these projects and their impact on the biosphere requires both a better understanding of the complexity and interconnectedness of microbial ecosystems and their connections to the rest of the biosphere, as well as a conservation policy that requires an adequate assessment of these connections prior to implementing such schemes. The importance of viruses in the worldwide marine ecology, particularly in the management of microbial populations, cannot be overstated. Despite the frequently unfavorable connection that viruses have in the medical setting, these findings demonstrate that viruses must be included in a microbial conservation effort (Suttle, 2005). As microbial communities are associated with biogeochemical cycles similarly some microbes also playing important role in regional- and local-scale cycles. Pollution may induce both regional and local decreases or changes in microbial populations. Large soil regions, forest ecosystems or marine and freshwater ecosystems represent areas of concern. Coral reefs that have a vast diversity of microbials are an excellent example (Cockell and Jones, 2009). Bourne et al. (2008) investigated microbiological changes on the Great Barrier Reef, Australia, focusing on the bacterial communities connected with coral bleaching. An increase in Vibrio species, including Vibrio spp., was seen in the water before the bleaching event occurred. The results show that bacterial populations are not immune to external forces like climate and environment. A variety of coral and other marine habitats will see dramatic changes in their microbial populations, which will in turn effect other creatures in the community.

7.5. Microbial Growth, Staining and Biochemical Tests

Most microbiologists define "mechanism" as the biochemical and physiological interactions that occur in a cell between genes and their products. To comprehend diversity in people and societies, this perspective must be broadened to include fitness. The ultimate judge of ecological success is fitness, which is a measure of one genotype's capacity to leave offspring in comparison to other types. The interactions between genotypes and between genotypes and their environment are the mechanisms causing variations in fitness (Kassen and Rainey, 2004). Consider an experiment reported over a century ago by Reverend William Henry Dallinger, a Wesleyan priest and president of the Royal Microscopical Society. Dallinger wondered in his president's presentation to the Society in 1887, "if it was feasible, by change of environment, in minute life-forms, whose life-cycle was relatively soon completed, to superinduce changes of an adaptive character, if the observations continued over a sufficiently long period." He began with a collection of microorganisms (dubbed monads) and grew them in a purpose-built incubator for seven years, steadily increasing the temperature until the apparatus was unintentionally destroyed (Gause, 2019). Dallinger returned his experimental populations, which had grown to temperatures much beyond their typical thermal tolerance limits, to the temperature at which he had begun the experiment just before the accident. None of them grew. He reasoned that the lines had adapted to rising temperatures through natural selection and, as a result, had lost their ability to grow at normal temperatures. It is worth noting that the processes used to explain the outcomes of this experiment are simple: Natural selection driven by rising temperatures leads to adaptability, but at the expense of not being able to survive at the ancestral temperature. Although understanding of molecular genetics, biochemistry, or physiological reasons of adaptation is desirable, it is not essential to interpret these data. The inability of the derived populations to grow in the ancestral environment shows that the changes that happened during the selection experiment were genetically determined. If Dallinger's experiment had not been destroyed, he could have gone on to determine the magnitude of improvement at elevated temperatures relative to the ancestral type, tested for and directly measured tradeoffs in ecological performance at various temperatures, and even documented diversity within evolving populations (its stability and changes in patterns of abundance through the course of the experiment). The astonishing fact is that, with relatively little extra development, Dallinger's approach from more than a century ago remains a relevant and potent way of studying some of biology's most fundamental challenges. Furthermore, such analyses can provide insight into the mechanisms of ecological and evolutionary change with the same strength and accuracy that a molecular microbiologist would anticipate when confronting an issue in gene regulation, for example. The experiment also underlines a second important issue for microbiologists interested in understanding the mechanisms of adaptability and diversity: evolution can be investigated on timelines that allow for experimentation and analysis. This should come as no surprise, given that microbial populations' quick generation rates and enormous population numbers are perfect for natural selection. Interestingly, Darwin himself passed up an opportunity to test his idea of adaptation by natural selection experimentally: Charles Darwin wrote to Dallinger on July 2, 1878, "I had no idea that you were attending to the mutation of lower species under different conditions of existence; and your discoveries, I have no doubt, will be highly intriguing and valuable." The fact that they were acclimated to temperatures yet progressively become adjusted much higher ones, as you describe, is rather astonishing. It explains why microorganisms grows in hot springs (Lack, 1983)

7.6. Industrial Applications

Microbes have proven themselves to be some of the best materials for biotechnology because of their varied metabolic abilities and recognized manipulation strategies. Microbial cultures are used to make alcohols, organic acids, and polymeric compounds in industrial operations (Baltz et al., 2010). In addition, fermented foods, a category of which contains a long history of people using various microbial cultures, are a prominent part of human culture (Giraffa, 2004). Additionally, many environmental biotechnology processes such as wastewater treatment, methanogenic digestion, bioremediation, and microbial fuel cells rely on naturally occurring microbial communities (Rittmann and McCarty, 2001). Investigations of the microbe communities participating in these processes have been the focus of scientists and engineers alike since the findings are not only informative about microbe functions but also about how ecosystems work and are influenced by individual organisms.

7.6.1. Microbes in waste management

Waste is not merely an afterthought; it is inextricably linked to the creation of spatial relationships at various mode of our daily life. We are all familiar with cleaning and squandering, and their products must be handled in some way or managed once disposed of. Yet, in many respects, the study of aftermath of our discarded products are just getting started. Anthropologists have had little to say about waste management until recently (Reno, 2015). This trend is thought to indicate a preference for "social" concepts over "individual" practices, which can be traced back to early epistemological divisions between science and technology. However, waste management techniques are fascinating in and of themselves. If classification standards influence how waste products are managed.

The waste management process is more than a by-product manufacture of a distinctly human urge for order; it is also a process actively involved in altering our ideals and imaginations. Appropriate waste management services are now regarded as a basic human right, an economic opportunity, and an ecological imperative in the governance of cities, industry, and refugee camps. Research on waste can also lead to the retrieval of valuable assets that would otherwise be discarded. The creative process of turning waste into something new is possible at every stage of the waste disposal process, including collection, cleaning, sorting, and disposal. A fruitful combination of human ingenuity, the vigor of waste materials, and the environment in which they are disposed makes an open-ended transformation feasible. If waste transmutation is ever employed, it could cause contamination, either literally or metaphorically. The former could mean having evidence left behind from the original carrier, and the latter could mean an undefined commodity: it could be an actual thing or just a component of its original material. Human activities generate large volumes of wastewater, much of which is treated biologically using activated sludge. These techniques use naturally occurring microbial communities, whose topologies fluctuate with changes in wastewater chemistry and environmental circumstances.

Thus, understanding microorganisms and their activity is critical to the effectiveness of these operations. Recent developments in nucleotide sequencing have ushered in a new era of molecular and genomic techniques in microbial ecology (Unno and Shinano, 2012; Kimura, 2007). Satoh et al. (2013) employed pyrosequencing to track bacterial population changes in laboratory activated sludge reactors (Satoh et al., 2013), collecting over 80,000 sequences. Even individual genus-level lineages like *Accumulibacter phosphatis* could be phylogenetically studied. Numerous microbes have been discovered that are capable of degrading and detoxifying a variety of contaminants, including xenobiotics and hazardous metals (Watanabe and Hamamura, 2003). Furthermore, recent advancements in molecular and genomic approaches, including as metagenomics and meta transcriptomics, have advanced our understanding of the metabolic pathways and genetic roles that underpin bioremediation processes (Desai et al., 2010). Interspecies interactions are critical for biodegradative microorganisms. For instance, new research indicates that the reductive de-chlorination of polychlorinated dipenzo-p-dioxins and the oxidative degradation of dechlorinated products occur concurrently in semi-anaerobic microbial microcosms, facilitating their total de-chlorination (Hiraishi et al., 2005; Kaiya et al., 2009).

7.6.2. Microbes in Therapeutics

Microbiology's contributions to advancements in the healthcare business, particularly in the pharmaceutical and medical industries, have resulted in significant discoveries, ranging from vaccinations to medical gadgets. The rise of the cosmetics industry coincided with advances in microbiology, which paved the way for the study of cosmetic microbiology. Pharmacists can find antimicrobial medicines by understanding microbiology and human cell processes, which will help to avoid an increasing number of communicable illnesses. Research has proven that microorganisms aren't simply dangerous diseases causing agents but may also be helpful to human health. Probiotics, or "live micro-organisms that, when given in appropriate numbers, impart a health benefit on the host," Clinical investigations have shown that some probiotics are quite effective. New sequencing tools have significantly enhanced the study of human microbiota and have established its significance (Berlec and Štrukelj, 2019). It is worthwhile to mention that biologically active compounds from food and commensal microorganisms can be expressed and made copies in the laboratory through genetic engineering (Gupta et al., 2014a). Engineering microorganisms for therapy can result in the improved microbial strains as well as selection of novel microbial strains using mutagenesis, genetic engineering, and synthetic biology (Berlec and Štrukelj, 2019). Probiotics that have been genetically modified can be used for the treatment of various inflammatory bowel illnesses like ulcerative colitis (Gupta et al., 2014b). Delivery of antigens for immunization is one of the most often investigated uses of genetically modified microorganisms. The use of the lactic acid bacteria *Lactococcus lactis* as a mucosal vaccine delivery vehicle has been studied by de Castro et al., (2018).

One essential feature in bacterial fitness is the production of antibiotic compound. Several antimicrobial chemicals are produced by different probiotic bacteria, including nitric oxide, bacteriocins, short chain fatty acids etc. Bacteriocins, produced by bacteria are antimicrobial compounds having varied host ranges. Food scientists can use the fact that food-grade lactic acid bacteria often generate bacteriocins to control or limit the development of bacterial species in food. Fermented food, for example, can benefit from the uses of these technologies in preservation and food safety, but also to create better flavours in food fermentation (Cotter et al., 2005). Nisin a bacteriocin has been examined as an antimicrobial compound and potential bio preservative used in industries processed foods preservation (Kumari et al., 2009).

The chemotherapeutic agents most crucial to treating cancer are from microbial origins. Nearly two-thirds of all existing anti-cancer drugs originated as naturally derived or modified natural products. Several compounds, including actinomycin D, bleomycin, anthracenones, enediynes (calcheamycin) etc. are some approved products (Gupta et al., 2014b).

Many hypocholesterolemic drugs can be obtained from microbes like Lovostatin, which was the first statin drug received FDA approval in 1987.

Lovastatin was isolated from *Monascus ruber* and *Aspergillus terreus.* Simvastatin is derivative of lovastatin. Atorvastatin, a synthetic variant of the natural statins, has been the industry leader for a long time (Deng, 2009; Monciardini et al., 2014)

7.6.3. Food and Dairy Production:

Fermented products have been the subject of microorganisms used in fermentation, as well as starter cultures, as well as a variety of technological, protective, and probiotic starters, and the issues of using them. Genetic and metabolomic characterization of microorganisms of technological impact on fermented products has also been completed. Additionally, pathogen and spoilage organism control and inhibition, as well as the relationship between technological procedures and microbiota, has been explored. Other ecosystems may include numerous species of microbes, but foods usually contain only a few numbers. Lactic acid bacteria (LAB) are a varied category of Gram-positive, catalase-negative bacteria that produce the acid. The European Food Safety Agency (EFSA) granted several food-related LABs the status of "Qualified Presumption of Safety" (QPS) or "General Recognized as Safe" (GRAS) by the US Food and Drug Administration. These bacteria and their metabolites are being used for a variety of bioactivities in addition to food preservation. LABs are particularly important in the generation of probiotics and starter cultures. In fact, LABs are collection of industrially important microorganisms that are employed in a wide range of food fermentations around the world, including dairy, wine, bread, vegetables, and more.

Preparing and testing a possible alternative to chemical preservatives for use in assuring the quality and safety of prepared foods with extended shelf life, fresh flavor, and minimal processing is underway now, with plans to expand the research in the near future. For alcoholic fermentation, *Saccharomyces*, and non-Saccharomyces yeasts, as well as *Bacillus spp.* for alkaline fermented foods, similar observations can be found. New starter cultures involving natural adaptation and evolution, such as direct selection of mutants with desired properties, adaptive laboratory evolution, and genetic methods other than genome sequencing of wild type strains, should be identified for guiding safety assessments and strain improvement activities. Soon, edible microbial biomass formed from bacteria, yeasts, filamentous fungi, or microalgae may be used to supplement traditional food sources. These bacteria are now recognized as a significant and valuable source of proteins, vitamins, and other bioactive substances. By the year 2100, the human population is expected to reach 9–12 billion people, and microbes could play a key role in ensuring a sustainable agricultural system.

Dairy products have been a part of the human diet since ancient times. These are high in calcium, vitamin D, protein, and other vital nutrients. As the demand for fermented food increases, dairy industries in many nations are contributing to economic growth. Fermented product production is commercially significant in several countries. Consumers can now choose from a wide range of fermented dairy products.

The dairy business is a great illustration of how bacteria, yeasts, molds, and viruses play a significant role in deciding the quality of the product. Control and elimination of unwanted microbes, as well as their planned introduction and exploitation, are issues that require specific attention. Milk's nutritional value makes it a desirable diet for humans and other young animals. These nutritious properties, however, allow for the growth of numerous bacteria, some of which produce negative alterations in milk and its products. Many factors influence the hygienic properties of milk during its production, processing, and distribution to consumers. Cheese production is an example of a dairy business that is heavily reliant on the desired enzymatic modifications generated by microorganisms. The conversion of milk ingredients by diverse species is largely responsible for the flavor and texture of cheese. Bacteria and mold species are purposely added to cheese during production, and much of the microbial activity that occurs during ripening is caused by microbial species that enter the milk by coincidence at various stages. Starter cultures, which are mixed bacterial cultures chosen for their potential to produce acid and flavor, are essential in the production of flavored butter. The parameters under which ideal flavor develops, as well as the relationships with bacteria, are well understood, and the modifications brought about by the starters are obviously desirable.

Yogurt, sour cream, and buttermilk are examples of fermented milks produced by distinct microorganism interventions in milk, cream, and skim milk, respectively. Each of these foods has a distinct microflora that contributes to its distinct flavor and texture. Using starter cultures, it is also feasible to ferment milks with desired features (for example, flavor and acid production).

Commercial conversion of specific milk ingredients into economically valuable products such as vitamins, solvents, and food additives are now possible. Milk ingredients that have little economic value or are frequently squandered, on the other hand, must be transformed to stably oxidized and non-obnoxious compounds before being discharged into the environment. Microbial activity is also responsible for the required modifications in the organic contents of dairy waste dung of sewage treatment in this case.

Microbes in milk or milk products are undesirable if they produce obnoxious flavor or texture and cause disease. It is critical that dairy microbiologists understand the factors that drive milk degradation and the procedures for preventing this sort of deterioration. Souring milk or cream is generally undesirable, although souring (the development of lactic acid) is required for the manufacturing of cheese or cultured milk products. Different microorganisms that thrive in dairy products cause discoloration, sliminess, ropiness, putrefaction, rancidity, gassiness, and a variety of other problems. Milk has occasionally been found to be a carrier of pathogenic microorganisms. These pathogens, on the other hand, can be effectively controlled. As dairy scientists develop new methods and products, they must be tested to guarantee that they are safe for customers.

Many laws and other regulations governing the production and handling of milk establish quantitative, qualitative, and microbiological criteria. These guidelines have demonstrated that, under ideal conditions, the number of microorganisms entering the product will not exceed levels, and that good treatment of these products will not allow growth beyond these levels. To some extent, the numbers of microorganisms permitted by these guidelines are typically well below the amount required to cause deterioration. In practice, this leads to a progressive improvement in microbiological standards for various dairy products. Because of this form of management, dairy companies' packaged milk is of high microbiological quality and thus safe for consumers.

Several studies have shown that germs carried through milk can cause diseases in both humans and animals. However, the existence of such microbes is not limited to dairy foods but can be found in several other meals as well. However, it is especially severe with dairy products because they are typically taken without being heated, resulting in the eradication of potentially hazardous microorganisms. As a result, understanding the sources of pathogens in dairy products, the environment under which they flourish, and ways for preventing or eliminating them is critical. Such expertise is crucial for dairy product manufacturers because of the obligation of creating only safe meals, and it is especially critical for dairy plant supervisors because inspection is the ultimate line of defense for the consumer. Although the manufacturer and supervisor are responsible for ensuring consumer safety, it is also preferable that consumers are well informed to determine whether a particular dairy product is satisfactory or not. Similarly, a farmer who uses milk or milk derivatives to feed his livestock should be aware of the risk of bringing diseases into his herds through such products.

Milk-borne infections such as tuberculosis, brucellosis, and typhoid fever have been reduced due to hygienic milk production procedures, safe milk handling and storage, and pasteurization. However, several milk-borne illnesses have occurred because of the consumption of raw milk or dairy products manufactured with milk that was not adequately pasteurized or handled improperly, resulting in post-processing contamination. In raw milk and other dairy products, the following bacterial infections are of particular concern: *Bacillus cereus, Campylobacter jejuni, E. coli, Listeria monocytogenes, Salmonella spp, Yersinia enterocolitica.*

It should be noted that molds, primarily *Aspergillus, Fusarium*, and *Penicillium* species, can grow in milk and dairy products. If the conditions are favorable, these molds may create mycotoxins, which can pose a health risk to consumers.

7.6.3.1. Milk Spoilage

Microorganisms generate unwanted changes in milk and are responsible for spoiling because of their growth or metabolic activities. The milk producer should be aware of the sources of microorganisms that cause quick changes, the settings that promote their growth, and strategies for avoiding their activity. The manufacturer of milk products must deal with issues similar to those faced by the producer of milk, as well as new ones, because butter, cheeses, and other dairy products are usually held for extended periods of time, which can lead to further deterioration of quality. The issue of spoiling is especially relevant with cheeses that require ripening since conditions must be favorable for the growth of some desirable microbes while simultaneously allowing the development of unwanted ones. Raw milk's initial microbiological quality is critical for the manufacture of high-grade dairy meals. Spoilage refers to the degradation of a food's texture, colour, odor, or flavor to the point that it is unfit for human consumption. The destruction of protein, carbs, and lipids by bacteria or their enzymes is a common cause of microbial deterioration of food. Psychrotrophs are the microorganisms that cause most of the deterioration in milk. Most psychrotrophs are eliminated by pasteurization, but others, such as *Pseudomonas fluorescens* and *Pseudomonas fragi*, can produce heat stable proteolytic and lipolytic extracellular enzymes capable of causing spoiling. *Bacillus, Clostridium, Corynebacterium, Arthrobacter, Lactobacillus, Microbacterium, Micrococcus*, and *Streptococcus* species and strains can survive pasteurization and develop at refrigerator temperatures, causing spoiling concerns in milk and milk products.

7.6.3.2. Developing Desirable Microorganisms in Dairy Products

Increasing the presence of beneficial bacteria in some milk products (Amadoro et al., 2018). The appealing flavor of dairy products such as butter, cheeses, and fermented milks, as well as the desired texture of many dairy products, are largely related to the development of specific microbes during manufacturing and ripening. Understanding the action of these organisms, the conditions that promote their growth, and so on is crucial and required if items of uniformly high quality are to be created and marketed. Manufacturing procedures for specific dairy products were created empirically even before the significance of microbes was identified, but knowledge of the activity of the main bacteria has resulted in numerous valuable advancements. The microflora of different dairy-based products is tabulated in table 7.1.

Table 7. 1. - Microbiological Requirements for milk and dairy Products[1]

S. no.	*Product description*	*Plate count (aerobic)*		*Coliform count*	
		Limit (cfu)		**Limit (cfu)**	
		m $	**M#**	**m $**	**M#**
1	Pasteurized Milk/Flavored Milk	30×103/ml	50×103/ml	Absent/0.1ml	-
2	Pasteurized Cream	50×103/g	75×103/g	10/g	-
3	Pasteurized Butter	25×103/g	50×103/g	Absent 0.1 g	20/g
4	Sweetened Condensed Milk	5×102/g	1×102/g	Absent 0.1 g	-
5	Paneer/ Chhana	15x104/g	35x104/g	10/g	1x102/g
6	Khoa	25x103/g	75x103/g	50/g	1x102/g

1 Food Safety and Standards (Food Products Standards and Food Additives) Regulations, 2011

$^{\$}$m= Represents an acceptable level and values above it is marginally acceptable

$^{\#}$M= A microbiological criterion which indicate unsatisfactory or potentially hazardous quality. Values above M are unacceptable in terms of the sampling plan and detection of one or more samples exceeding this level would be cause for rejection of the lot and will attract prosecution by the concerned food safety authorities.

Butter and powdered infant foods are popular dairy products and accepted throughout the world. The microbiological standard with respect to yeast and mold count of butter and mesophillic aerobic bacteria in India is tabulated in table 7.2 and table 7.3.

Table 7.2. - Microbiological Standards (Bureau of Indian Standards) of Butter

Yeast & Mold count/ml	*Quantity*
< 20	Good
21-50	Fair
51-100	Poor
> 100	Very Poor

Table 7. 3. - Microbiological quality of Powdered Infant formula products

Powdered Infant formula products	(n)	(c)	(m)	(M)
Mesophilic Aerobic Bacteria	5	2	500/g	5000/g
Enterobacteriaceae[2]	10	2	0/10g	-

n = number of sample units

c = the number of sample units allowed to exceed m

m = the acceptable microbiological limit

M = the limit which must not be exceeded

A list of industrial enzymes produced by various microorganisms are enlisted in table 7.4.

---

2 *Codex proposed a 2-class sampling plan for Enterobacteriaceae on the basis that a 3 class sampling plan would not be practical analytically given the low levels of Enterobacteriaceae that occur when stringent hygiene conditions are maintained. This criterion assumes that:*

- *the product is sufficiently homogenous so that high level contaminations will fail (more than two samples would exceed "m")*
- *in practice, positives would not normally be found if strict hygiene measures are in place. If occasional positives are found, the manufacturer would take appropriate actions.*

Table 7. 4. - List of Industrial enzymes and microorganisms producing them

Enzyme	Activity	Microorganisms
Protease	Proteolytic activity	*Pseudomonas sp; Clostridium sp; Rhizopus sp; Penicillium sp; Aspergillus sp*
Amylase	Starch hydrolyzing activity	*Bacillus sp.; Geobacillus sp*
Xylanase	Xylan polymer hydrolyzing activity	*Thermoactinomyces thalophilus; Bacillus sp.; Humicola insolens*
Cellulase	Cellulose polymer hydrolyzing activity	*Bacillus subtilis; Acinetobacter junii; Rhodothermus marinus;*

7.6.4 Bioremediation

Bioremediation is a biological technique for recycling wastes into a form that other species may utilize and reuse. The entire world is currently dealing with various forms of environmental contamination. Microorganisms are a critical component of a key alternative strategy for overcoming obstacles. Microorganisms can survive in every location on the biosphere due to their incredible metabolic activity, and they can thrive in a wide range of environmental conditions. Microorganisms have a wide range of nutritional capacities, which makes them ideal for bioremediation of environmental contaminants (Abatenh et al., 2017). Through the all-inclusive and action of microorganisms, bioremediation is heavily involved in the degradation, eradication, immobilization, or detoxification of various chemical wastes and physical dangerous chemicals from the surrounding. Degrading and changing pollutants such as hydrocarbons, oil, heavy metals, pesticides, dyes, and other chemicals is the primary principle. Because it is carried out in an enzymatic manner by metabolizing, it has a significant contribution to the solution of many environmental problems. The pace of deterioration is determined by two types of factors: biotic and abiotic environments. Various tactics and strategies are currently being used in the domain in various parts of the world. Bio-augmentation, bio-stimulation, bioventing, and bio-attenuation are only a few examples. Because each bioremediation technology has its own unique application, each has its own set of benefits and drawbacks.

Biodegradation is a very profitable and appealing technology for cleaning, controlling, and restoring polluted environments through microbial activity. The rate at which undesired waste chemicals degrade is determined by competition with biological agents, insufficient food supply, unpleasant external abiotic conditions (e.g.: aeration, moisture, pH, temperature), and limited pollutant bioavailability. Because of these characteristics, biodegradation under natural conditions is less successful, resulting in less beneficial results. Bioremediation is only successful when the environment allows for microbial growth and activity. Therefore, bioremediation has been employed in a variety of locations around the world with variable degrees of effectiveness. The advantages outnumber the cons in most cases, as seen by the growing number of sites that use this technology and its increasing popularity over time. Generally, different species are investigated from various locations and found to be efficient control mechanisms.

Microorganism like *Myrothecium roridum* (IM 6482), *Pycnoporus sanguineous, Phanerochaete chrysosporium, Trametes trogii* were successfully exploited for removal of industrial dyes. (Jasińska et al., 2015; Yan et al., 2014). *Acinetobacter baumanii, Bacillus cereus, Exiguobacterium indicum* were also reported for removal of azo dyes from industrial effluents (Kumar et al., 2016). Several studies have demonstrated the inherent ability of many species (both procaryotes and eukaryotes) to bio-sorb heavy metal ions. *Cunninghamella elegans, Pseudomonas fluorescens* and *Pseudomonas aeruginosa* were reported for removal of heavy metals such as lead, iron, zinc, cupper etc. (Tigni et al., 2010; Paranthaman and Karthikeyan, 2015). Recent research has shown the capacity to remove heavy metals from environments by certain fungi (such as *Aspergillus sp* and *Penicillium sp*) and some yeasts (*S. cerevisiae*, for example). Heavy metals have also been removed from water bodies using the bacteria *Escherichia coli, Bacillus subtilis, Saccharomyces boulardii, Enterococcus faecium,* and *Staphylococcus aureus* (Bhakta et al., 2014).

7.6.5 Biofuel Production

Biofuels are fuels made from organic matter, or lignocellulosic biomass, such as corn or sugar, vegetable oils, or waste feedstocks. Bioethanol and biodiesel are two known and popular examples of biofuel. The consumption of organic substrates by a microbe and subsequent usage in metabolic processes produces valuable compounds that can be utilized as a fuel to generate energy. For biofuel synthesis, the selection of microorganisms, substrates, and production techniques is critical. For biofuel production, the microorganisms have a specialized metabolic route and various types of catalytic enzymes. A common example of this is the direct decarboxylation of pyruvate in *Saccharomyces cerevisiae*, which produces ethanol, but the same reaction in Escherichia coli is activated by CoA, which reduces to ethanol (Bhatia et al., 2018).

The fuel production pathway must be introduced into a microbial host to attain the required to be advanced biofuel. Several factors influence the choice of microbial host, including the carbon source (e.g., carbon dioxide, cellulose, lignin, methane etc.), the toxic effects of the substrate or the fuel itself, as well as the processing parameters required to transform the substrate and produce the fuel. Heterologous routes for producing improved biofuels have been transferred to a wide range of hosts, allowing for the utilization of a wider range of substrates and growing conditions. Beyond the basic work in *E. coli* and *S. cerevisiae*, iso-butanol synthesis has been engineered into a wide range of hosts. *Corynebacterium glutamicum*, an industrial actinobacterium with increased tolerance to iso-butanol and a native high throughput synthesis of branched chain amino acids, precursors to iso-butanol, has had the iso-butanol pathway engineered into it (Blombach et al., 2011). New research on biodiesel (fatty acid methyl esters) made from microalgae, yeast, and bacteria is focusing on its potential to replace petroleum fuel. There are several energy studies for algae-based biofuel production routes, however these often do not include energy integration alternatives such as utilizing industrial excess heat as a heat source (Ubando et al., 2019). The algal genera which produce maximum amount of biofuel are *Chlorella sp, Chlamydomonas sp, Nannochloropsis sp, Scenedesmus sp, Spirulina sp.*

7.7. Conclusion

Microbial products can range from big components like proteins, nucleic acids, and carbohydrate polymers to microscopic molecules like metabolites that are needed for vegetative growth (primary metabolites) and those that are nonessential (secondary metabolites). We rely on microorganisms for numerous reasons, but the main one is that they provide us with valuable goods. One reason for microorganisms' production of compounds is their large surface area compared to volume, allowing them to rapidly uptake nutrients and process reactions. Another reason is their wide variety of reactions and the ability to be moved from nature to the laboratory flask or factory fermenter, where they can grow on inexpensive carbon and nitrogen sources to produce valuable compounds. (iv) a significantly increased production up to thousands-fold, by means of genetic manipulation, both in vivo and in vitro; (v) simplicity of screening procedures that allows thousands of cultures to be analyzed in a short time; and (vi) a broad range of enzyme variants that produce the same reaction, giving the operator more flexibility in reactor operating conditions.

Reference

Allison, S. D. & Martiny, J. B. (2008). Resistance, resilience, and redundancy in microbial communities. Proceedings of the National Academy of Sciences, 105(Supplement 1), 11512-11519.

Amadoro, C., Rossi, F., Pallotta, M. L., Gasperi, M. & Colavita, G. (2018). Traditional dairy products can supply beneficial microorganisms able to survive in the gastrointestinal tract. LWT, 93, 376-383.

Abatenh, E., Gizaw, B., Tsegaye, Z. & Wassie, M. (2017). The role of microorganisms in bioremediation-A review. Open Journal of Environmental Biology, 2(1), 038-046.

Arrigo, K. R. (2005). Marine microorganisms and global nutrient cycles. Nature, 437(7057), 349-355.

Baltz, R. H., Demain, A. L. & Davies, J. E. (Eds.). (2010). Manual of industrial microbiology and biotechnology. American Society for Microbiology Press.

Berlec, A. & Štrukelj, B. (2019). Engineering microbes for therapy. Frontiers in microbiology, 10, 629.

Bhakta, J. N., Munekage, Y., Ohnishi, K., Jana, B. B. & Balcazar, J. L. (2014). Isolation and characterization of cadmium-and arsenic-absorbing bacteria for bioremediation. Water, Air, & Soil Pollution, 225(10), 1-10.

Bhatia, L., Singh, A., Chandel, A. & Singh, O. M. (2018). Biotechnological Advancements in cellulosic ethanol production. O. V. Singh and A. K. Chandel (eds.), Sustainable Biotechnology- Enzymatic Resources of Renewable Energy, https://doi.org/10.1007/978-3-319-95480-6.

Blombach, B., Riester, T., Wieschalka, S., Ziert, C., Youn, J. W., Wendisch, V. F. & Eikmanns, B. J. (2011). Corynebacterium glutamicum tailored for efficient isobutanol production. Applied and environmental microbiology, 77(10), 3300-3310.

Bourne, D., Iida, Y., Uthicke, S., & Smith-Keune, C. (2008). Changes in coral-associated microbial communities during a bleaching event. The ISME journal, 2(4), 350-363.

Boyd, P. W., Jickells, T., Law, C. S., Blain, S., Boyle, E. A., Buesseler, K. O. & Watson, A. J. (2007). Mesoscale iron enrichment experiments 1993-2005: synthesis and future directions. science, 315(5812), 612-617.

Bull, A.T. (Ed.). (2004). Microbial diversity and bioprospecting (No. 660.62 M53). Washington, DC: ASM press.

Cockell, C. S. & Jones, H. L. (2009). Advancing the case for microbial conservation. Oryx, 43(4), 520-526.

Costa, S., Lopes, I., Proença, D.N., Ribeiro, R. & Morais, P.V. (2016). Diversity of cutaneous microbiome of Pelophylax perezi populations inhabiting different environments. Science of the Total Environment, 572, 995-1004.

Cotter, P. D., Hill, C. & Ross, R. P. (2005). Bacteriocins: developing innate immunity for food. Nature Reviews Microbiology, 3(10), 777-788.

Challis, G. L. (2008). Mining microbial genomes for new natural products and biosynthetic pathways. Microbiology, 154(6), 1555-1569.

Daniel, H. M. & Prasad, G. S. (2010). The role of culture collections as an interface between providers and users: the example of yeasts. Research in microbiology, 161(6), 488-496.

Demain, A. L. & Adrio, J. L. (2008). Contributions of microorganisms to industrial biology. Molecular biotechnology, 38(1), 41-55.

Deng, R. (2009). Food and food supplements with hypocholesterolemic effects. Recent patents on food, nutrition & agriculture, 1(1), 15-24.

de Castro, C. P., Drumond, M. M., Batista, V. L., Nunes, A., Mancha-Agresti, P. & Azevedo, V. (2018). Vector development timeline for mucosal vaccination and treatment of disease using Lactococcus lactis and design approaches of next generation food grade plasmids. Frontiers in microbiology, 9, 1805.

Desai, C., Pathak, H. & Madamwar, D. (2010). Advances in molecular and "-omics" technologies to gauge microbial communities and bioremediation at xenobiotic/anthropogen contaminated sites. Bioresource Technology, 101(6), 1558-1569.

Fields, M. W., Yan, T., Rhee, S. K., Carroll, S. L., Jardine, P. M., Watson, D. B. & Zhou, J. (2005). Impacts on microbial communities and cultivable isolates from groundwater contaminated with high levels of nitric acid–uranium waste. FEMS Microbiology Ecology, 53(3), 417-428.

Gause, G. F. (2019). The Struggle for Existence: A Classic of Mathematical Biology and Ecology. Courier Dover Publications.

Gilbert, J. A. & Neufeld, J. D. (2014). Life in a world without microbes. PLoS biology, 12(12), e1002020.

Giraffa, G. (2004). Studying the dynamics of microbial populations during food fermentation. FEMS Microbiology Reviews, 28(2), 251-260.

Gupta, C., Prakash, D. & Gupta, S. (2014a). Natural useful therapeutic products from microbes. J Microbiol Exp, 1(1), 00006.

Gupta, C., Prakash, D., Garg, A. P. & Gupta, S. (2014b). Nutraceuticals from microbes. Phytochemicals of nutraceutical importance. Wallingford: CABI International Publishers, 79-102.

Gupta, A., Gupta, R. & Singh, R. L. (2017). Microbes and environment. In Principles and applications of environmental biotechnology for a sustainable future (pp. 43-84). Springer, Singapore.

Gutleben, J., Chaib De Mares, M., Van Elsas, J. D., Smidt, H., Overmann, J. & Sipkema, D. (2018). The multi-omics promise in context: from sequence to microbial isolate. Critical reviews in microbiology, 44(2), 212-229.

Hiraishi, A., Kaiya, S., Miyakoda, H. & Futamata, H. (2005). Biotransformation of polychlorinated dioxins and microbial community dynamics in sediment microcosms at different contamination levels. Microbes and Environments, 20(4), 227-242.

Hayat, R., Ali, S., Amara, U., Khalid, R. & Ahmed, I. (2010). Soil beneficial bacteria and their role in plant growth promotion: a review. Annals of microbiology, 60(4), 579-598.

Jasińska, A., Paraszkiewicz, K., Sip, A. & Długoński, J. (2015). Malachite green decolorization by the filamentous fungus Myrothecium roridum–mechanistic study and process optimization. Bioresource technology, 194, 43-48.

Kaiya, S., Utsunomiya, S., Suzuki, S., Yoshida, N., Futamata, H., Yamada, T. & Hiraishi, A. (2009). Isolation and functional gene analyses of aromatic-hydrocarbon-degrading bacteria from a polychlorinated-dioxin-dechlorinating process. Microbes and environments, 1112010337-1112010337.

Kassen, R. & Rainey, P. B. (2004). The ecology and genetics of microbial diversity. Annu. Rev. Microbiol., 58, 207-231.

Kimura, N. (2007). Metagenomics: access to unculturable microbes in the environment. Microbes and Environments, 21(4), 201-215.

Knutie, S. A. (2018). Relationships among introduced parasites, host defenses, and gut microbiota of Galapagos birds. Ecosphere, 9(5), e02286.

Kostka, J. E., Prakash, O., Overholt, W. A., Green, S. J., Freyer, G., Canion, A. & Huettel, M. (2011). Hydrocarbon-degrading bacteria and the bacterial community response in Gulf of Mexico beach sands impacted by the Deepwater Horizon oil spill. Applied and environmental microbiology, 77(22), 7962-7974.

Konopka, A. (2009). What is microbial community ecology. The ISME journal, 3(11), 1223-1230.

Kumar, S., Chaurasia, P. & Kumar, A. (2016). Isolation and characterization of microbial strains from textile industry effluents of Bhilwara, India: analysis with bioremediation. J Chem Pharm Res, 8(4), 143-150.

Kumari, A., Makeen, K., Garg, A. P., Marotta, F. & Gupta, C. (2009). Effect of the bacteriocin produced by Lactococcus lactis subsp. lactis CCSUB202, on mode of action of Lactococcus lactis subsp. lactis MTCC3038. International Journal of Probiotics and Prebiotics, 4(3), 205-210.

Lack, D. (1983). Darwin's finches. CUP Archive.

Lal, R., Pandey, G., Sharma, P., Kumari, K., Malhotra, S., Pandey, R. & Oakeshott, J. G. (2010). Biochemistry of microbial degradation of hexachlorocyclohexane and prospects for bioremediation. Microbiology and molecular biology reviews, 74(1), 58-80.

Monciardini, P., Iorio, M., Maffioli, S., Sosio, M. & Donadio, S. (2014). Discovering new bioactive molecules from microbial sources. Microbial biotechnology, 7(3), 209-220.

Paranthaman, S. R. & Karthikeyan, B. (2015). Bioremediation of heavy metal in paper mill effluent using Pseudomonas spp. Int J Microbiol, 1, 1-5.

Prakash, O., Nimonkar, Y. & Shouche, Y. S. (2013). Practice and prospects of microbial preservation. FEMS microbiology letters, 339(1), 1-9.

Ramsby, B. D., Hoogenboom, M. O., Whalan, S. & Webster, N. S. (2018). Elevated seawater temperature disrupts the microbiome of an ecologically important bioeroding sponge. Molecular Ecology, 27(8), 2124-2137.

Reno, J. (2015). Waste and waste management. Annual Review of Anthropology, 44, 557-572.

Rittmann, B.E. & McCarty, P.L. (2001). Environmental biotechnology. McGraw-Hill, New York.

Senni, K., Pereira, J., Gueniche, F., Delbarre-Ladrat, C., Sinquin, C., Ratiskol, J. ... & Colliec-Jouault, S. (2011). Marine polysaccharides: a source of bioactive molecules for cell therapy and tissue engineering. Marine drugs, 9(9), 1664-1681.

Singh, B. K., Bardgett, R. D., Smith, P. & Reay, D. S. (2010). Microorganisms and climate change: terrestrial feedbacks and mitigation options. Nature Reviews Microbiology, 8(11), 779-790.

Staley, J. T., Castenholz, R. W., Colwell, R. R., Holt, J. G., Kane, M. D., Pace, N. R. ... & Tiedje, J. M. (1997). The microbial world. The Foundation of the Biosphere, 32.

Sitaraman, R. (2018). Prokaryotic horizontal gene transfer within the human holobiont: ecological-evolutionary inferences, implications and possibilities. Microbiome, 6(1), 1-14.

Smith, D. (2003). Culture collections over the world. International Microbiology, 6(2), 95-100.

Sharma, A., & Shouche, Y. (2014). Microbial culture collection (MCC) and international depositary authority (IDA) at national centre for cell science, Pune. Indian journal of microbiology, 54(2), 129-133.

Tedeschi, R. & De Paoli, P. (2011). Collection and preservation of frozen microorganisms. In Methods in biobanking (pp. 313-326). Humana Press, Totowa, NJ.

Suttle, C. A. (2005). Viruses in the sea. Nature, 437(7057), 356-361.

Satoh, H., Oshima, K., Suda, W., Ranasinghe, P., Li, N., Gunawardana, E. G. W. ... & Mino, T. (2013). Bacterial population dynamics in a laboratory activated sludge reactor monitored by pyrosequencing of 16S rRNA. Microbes and environments, 28(1), 65-70.

Ubando, A. T., Rivera, D. R. T., Chen, W. H. & Culaba, A. B. (2019). A comprehensive review of life cycle assessment (LCA) of microalgal and lignocellulosic bioenergy products from thermochemical processes. Bioresource technology, 291, 121837.

Unno, Y. & Shinano, T. (2012). Metagenomic analysis of the rhizosphere soil microbiome with respect to phytic acid utilization. Microbes and environments, ME12181.

Watanabe, K. & Hamamura, N. (2003). Molecular and physiological approaches to understanding the ecology of pollutant degradation. Current Opinion in Biotechnology, 14(3), 289-295.

Yan, J., Niu, J., Chen, D., Chen, Y. & Irbis, C. (2014). Screening of Trametes strains for efficient decolorization of malachite green at high temperatures and ionic concentrations. International Biodeterioration & Biodegradation, 87, 109-115.

Zhang, J., Liu, Y. X., Guo, X., Qin, Y., Garrido-Oter, R., Schulze-Lefert, P. & Bai, Y. (2021). High-throughput cultivation and identification of bacteria from the plant root microbiota. Nature Protocols, 16(2), 988-1012.

Zogg, G. P., Zak, D. R., Ringelberg, D. B., White, D. C., MacDonald, N. W. & Pregitzer, K. S. (1997). Compositional and functional shifts in microbial communities due to soil warming. Soil Science Society of America Journal, 61(2), 475-481.

Chapter-8

BIOREFINERIES FOR BIOFUELS AND PLATFORM CHEMICALS

Latika Bhatia[1]* and Dilip Kumar Sahu

Department of Microbiology & Bioinformatics, Atal Bihari Vajpayee University, Bilaspur, Chhattisgarh, India.

**Corresponding author Email: latikabhatia1@yahoo.co.uk*

Abstract

Until recently, fossil fuel-based resources were the base for generation of utmost of energy and industrially formed chemicals. Presently, there has been a shift in this conventional practice and the exploration of more sustainable and eco-friendly feedstocks are perused to generate fuels and chemicals, as the fossil fuels are depleting at a high pace and their impact on environment is adverse. Biomass are a sustainable and an alternative renewable resource that have been identified as a substitute to lessen the dependency of this sector on fossil fuel-based resources and to relieve their impressions on environment. Recent years have witnessed intensive research on exploring various lignocellulosic resources for industrial production of various biofuels and platform chemicals. Being abundant and renewable resource, these lignocellulosic biomasses can successfully replace fossil fuels-based resources. This chapter provides a deep insight on various aspects of biorefineries, their types and classification along with strategies employed in production of various biofuels and platform chemicals and the stage for commercialization.

Keywords: *Biomass, Biorefinery, Platform chemicals, Lignocellulosic Biomass, Bio-economy, Fermentation*

8.1. Introduction

World population is growing continuously and is demanding more material, energy and food. According to latest data of United nation, present global population is 7.3 billion and is presumed to rise to 9.7 billion by 2050. The escalating demands food, materials, and energy will be sufficed through competent usage of natural resources. By the year 2050, food demands are expected to rise from 59 to 98% (Valin et al., 2013). Presently, crude petroleum is the chief energy source and various chemicals and has been tagged as non-eco-friendly due to many reasons (Mishra et al., 2019). Crude petroleum is not only non-renewable, but are also the prime cause of climatic changes and enhances emissions of greenhouse gases. Therefore, the time demands to explore the sustainable substitute of crude petroleum (Chandel et al., 2018). Amongst the ensuring substitutions of gasoline or crude petroleum, agricultural crops and their products, lignocellulosic biomasses like agroforest residues, energy crops, and grasses, etc., and algal biomass have depicted outstanding advancement and could be hopeful environment friendly base products (Bhatia et al., 2020).

Presently, the mainstream of chemicals and energy are obtained from non-renewable fossil fuel-based resources. These depleting fossil fuels suffices 80% of energy and 90% of chemicals needs of the commercial markets and global societies (Fernando et al., 2006). The emanation of detrimental greenhouse gases and lethal materials are the consequences of using these fossil fuels for large scale production of energy and chemicals, thereby implying the negative impact on environment. The need for energy and chemicals is escalating considerably with the hike in global population. Along with a growth in global population, the equivalent amount of waste produced is also progressively intensifying (Takkellapati et al., 2018).

In this present scenario, it's an obligation to explore and identify alternative resources to generate energy and chemicals. This will not only resolve the environmental issues, but would also suffice the escalating necessity for chemicals and energy. Biomasses are the renewable resources that have been identified as a source to produce fuels and chemicals thereby, also addressing the issues of waste generation. Numerous waste sources like agro-wastes (corn stover, rice husk, peels and sugarcane bagasse), industrial waste (pulp and paper mills) and municipal solid waste (MSW), (Fig. 8.1) are the potential alternative sources of biomass along with the natural resources. An integrated waste management approach supports the valorization of waste biomass for resource recovery (Bhatia and Johri, 2016; Bhatia and Johri, 2017). A biorefinery, presents a sustainable approach for the production of platform chemicals by translating harvested and waste biomass thereby, utilizing renewable biomasses as a feedstock resource (Esposito and Antonietti, 2015). The current decade has witnessed a bio-refinery methodology sourcing agro-industrial waste matter as feedstock becoming an interesting ideal area in biotechnology. There are various reasons behind this trend, chief among them are depleting petroleum resources, risk of emission of greenhouse gases and swinging market worth of crude oil. A biorefinery supports sustainable growth, and is an eco-friendly approach as it reduces overall emissions of green-house gases (NREL) and air toxics (Hakkim et al., 2020). Ecological alteration of biomass into an extensive array of valued chemicals, fuels, and materials is the ultimate aim of a biorefinery (Liu et al., 2019).

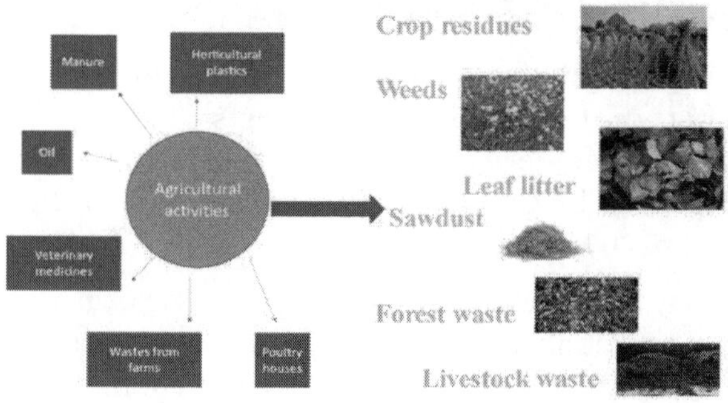

Figure 8.1. – Various agrowastes generated out of various agricultural activities

8.2. Biorefinery

It was during late 1990s, there took the evolution of the concept of biorefinery. Various organizations have defied biorefinery in unique way. As per the US Department of Energy (USDoE), biorefinery is a consolidate idea of processing plant where a wide array of valuable products is obtained from conversion or extraction of biomass feedstock (Kamm et al., 2006). In accordance with National Renewable Energy Laboratory (NERL), a biorefinery is an integration of biomass alteration procedures and equipment to generate power, chemicals, and fuels (NREL). According to the International Energy Agency (IEA) Bioenergy Task 42, a biorefinery is the supportable process of biomass conversion into a variety of vendible products viz. food, feed, chemicals, materials and energy (fuels, power, heat). In a nut shell, thus, a biorefinery can be a procedure, a plant, a facility or a collection of amenities for the biomass translation (de Jong et al., 2013). Additionally, a biorefinery can be integrated to a refinery based on petroleum. In petroleum-based refinery, energy and chemicals are produced from oil and natural gas (fossil-based resources), whereas in a biorefinery, energy and chemicals are produced from biomass feedstock (de Jong and Jungmeier, 2015). Figure 8.2 depicts the various advantages of biorefineries

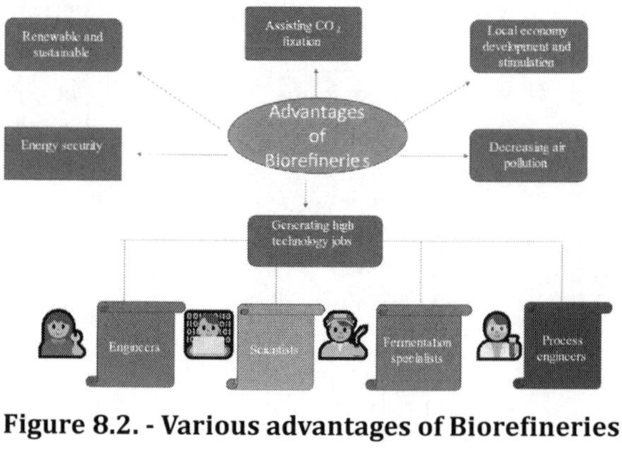

Figure 8.2. - Various advantages of Biorefineries

8.3. Classification of Biorefineries

Biorefinery supports the production of chemicals, power and fuel from various biomasses by incorporating steps that converts biomass along with various facilities, equipment/machinery and procedural configurations. With no doubts, biorefinery concepts can be contemplated equivalent to conventional petroleum refineries (Cherubini, 2010).

There are many bases on which biorefineries can be classified, thereby expanding its various concepts. First base can be the conversion processes employed, that classifies biorefineries into thermochemical, biochemical, two-platform process. On the basis of status of technology executed, biorefineries can be either conventional or advanced. The type of feedstocks employed and the form of intermediates generated (syngas or sugar) also formulate the basis to classify biorefineries. Thus, biorefineries are classified into three different types (Table 8.1) (Kamm and Kamm, 2004; Clark and Deswarte, 2008).

Table 8.1. - Types of biorefineries

S. No.	*Biorefinery Type*	*Important features*	*Examples*		*References*
			Substrate	**Product**	
1.	Phase I Biorefinery	• employs single feedstock material • has fixed processing competence • yields a single primary product.	vegetable oil	biodiesel	(Naik et al., 2010).
			pulp and paper mills	biodiesel	
			corn grain	ethanol	
2.	Phase II Biorefinery	• uses only one feedstock • capable of producing various products	starch	various chemicals	
			cereal grains	multiple carbohydrate derivatives	
			cereal grains	bioethanol	
3.	Phase III Biorefineries	• advanced biorefineries • can utilize numerous feedstocks, processing technologies • generates multiple types of products.	starch	various chemicals	(Clark and Deswarte, 2008).
			grass, green crops, or plants	Fuel	

There are four classes of Phase III Biorefinery systems-

- ☆ **A whole-crop biorefinery** – This type of biorefinery employs the whole crop, such as cereal grains, as feedstock. The first step involves the separation of cereal grain into corn and straw. Thereafter, corn is converted into starch, which upon hydrolysis yields glucose that acts as a substrate or feedstock for producing numerous chemicals. The cellulosic straw goes to lignocellulosic biorefinery for its further procession.
- ☆ **Green biorefinery** - This type of biorefinery employs natural biomass like grass, green crops, or plants as a feedstock.
- ☆ **Lignocellulosic biorefinery** - the chief feedstock of this type of biorefinery is a dry lignocellulosic biomass, like cellulose-based biomass.
- ☆ Newly, National Renewable Energy Laboratory (NREL) has presented the two-platform concept biorefinery - This type of biorefinery was recently introduced by NERL and comprises of two diverse kinds of platforms, viz. sugar platform and syngas platform.
    - ★ The sugar platform utilizes biochemical conversion procedures like fermentation of sugars gained from biomass.
    - ★ The syngas platform relies on thermochemical procedures like gasification of biomass to generate syngas. Energy and chemicals are produced as the end products by both these platforms.

Based on the chemical composition (nature) of feedstock employed in the production, a biorefinery can be further classified into three types (Octave and Thomas, 2009).

- ☆ **Triglyceride biorefinery (TGB)** – This type of biorefinery employs feedstocks like animal fats, vegetable oils, waste cooking oil and algae derived oil. A considerable attention has been received by this biorefinery because of scientific progressions made in the biofuels production. These triglycerides can be efficiently transformed into a biofuel through a transesterification reaction with methanol in the presence of a catalyst (acid, base, or enzyme). Glycerol is produced as a byproduct in this process. Among the various platform chemicals, glycerol and its derivatives have huge market potential, that adds a marketable value to this process.
- ☆ **Sugar and starchy biorefinery (SSB)** – There are many sophisticated and -executed fermentation processes that have potential to convert sugar and starch (feedstocks of this type of biorefinery) into ethanol (Serrano-Ruiz and Dumesic, 2011). Sugar beet, wheat, rice, sugarcane, maize, and corn are the feedstocks for SSB. During fermentation, starch or sugar polymers are first enzymatically degraded to the constituent monomeric sugars, after which fermenting yeast like Saccharomyces cerevisiae generates ethanol at an ambient temperature.

☆ **Lignocellulosic biorefinery (LCB)** - A lignocellulosic-based feedstock (LCF) is utilized in lignocellulosic biorefinery. LCF have potential to produce an extensive spectrum of products via innumerable processing methods. Cellulose, hemicellulose and lignin are the chief constituents of LCF and their composition varies in different feedstocks. Important examples of lignocellulosic feedstocks (LCFs) are wood, husk, straw, shells, baggase, grasses, etc. (Alonso et al., 2010).

Biorefinery can also be categorized into two types: First is conventional biorefinery and second is modern biorefineries/advanced biorefineries. Former involves the production of beer, wine, paper and vegetable oil, the practice that is going through centuries. Later, the mid-nineteenth century witnessed the revolutions in food production (sugar, potato starch) that permitted the production of products based on wheat, corn and starch in the twentieth century. Whereas the main aim of biorefineries is the exploration of utilization of algal and lignocellulosic biomass into fuel, energy, nutraceuticals and value-added chemicals. Modern biorefineries are still in their advancement phase and have the prospective to suffice the escalating future need for food/feed and energy (Bhatia et al., 2020).

IEA Bioenergy Task 42 developed an International Energy Agency (IEA) biorefinery classification in the year 2008. The criteria that played an important role in developing this classification were feedstocks, products, biorefinery platform, and processes (de Jong et al., 2013). Table 8.2 depicts this IEA biorefinery classification system.

Table 8.2. - IEA biorefinery classification system

S. No	*Biorefinery type*	*Feedstock used*	*Product*
1.	One-platform C_6 sugar biorefinery	corn crops	animal feed and bioethanol
2.	One-platform syngas biorefinery	lignocellulosic residues	chemicals and biofuels
3.	Two-platform (biogas and organic juice) biorefinery	grasses	biomethane, biomaterials, chemicals, (fiber products) and fertilizer
4.	Four-platform (lignin/syngas, C5/C6 sugar) biorefinery	lignocellulosic crop such as switchgrass	bioethanol, liquid biofuel and animal feed

8.4. Industrial Production of Biofuel

8.4.1. Biohydrogen

Hydrogen is a high energy density fuel that generates less pollutants on its combustion. These features make hydrogen a striking fuel in midst of numerous biofuels (Hallenbeck and Ghosh, 2009). Chemical industry employs hydrogen as a feedstock for detoxifying numerous water pollutants. Presently, these needs are sufficed by producing hydrogen via reformation of fossil fuel, a process that has a huge negative impact on environment, due to its greenhouse gas footprint (Redwood et al., 2009). Contrary to this, biological mode of hydrogen production is a sustainable approach and provides a possibility to produce hydrogen through carbon neutral approaches (Lee et al., 2010). Variety of biomass is utilized now a days for producing hydrogen, that includes agricultural crops and agro residues, lignocellulosic biomass, waste of food processing, aquatic plants and algae, and municipal effluents. As algae is known potential producer of hydrogen, substantial research has been done on genetic engineering of this organism. Techniques of genetic engineering are utilized to recognize appropriate bioenergy genes to augment the targeted biofuel production (Beer et al., 2009).

There are three procedures to produce biohydrogen (1) photosynthesis, (2) fermentation and (3) microbial electrolysis cells (MEC). The process of photosynthesis drives its energy from light source is mediated through hydrogenase enzyme that splits the water molecules to acquire hydrogen and oxygen. However, this process poses a limitation, that the activity of hydrogenase enzyme gets affected in presence of oxygen as it is low oxygen tolerant. Fermentative mode involves dark fermentation of organic substrates to produce biohydrogen. Protons are utilized in this process as electron acceptors. The best part of this process is that it allows the utilization of complex organic substrates thereby serving as an alternative approach to decompose organic substrates from municipal wastes or other organic wastes. Magnitude of hydrogen production in terms of relative hydrogen yield is far better via fermentation mode when compared with other modes of its production. There is also immense scope in improving the hydrogen yield via fermentative approach.

Nonetheless the theoretical utmost production of substrate electron corresponding to biohydrogen is 25%, only 17% conversion is truly accomplished. Feasible approaches to intensify the percentage alteration of substrate to biohydrogen comprise pairing the fermentation process with other routes like microbial electrolysis cells (MEC). According to recent publications, the association of dark fermentation with MEC can upgrade the hydrogen yield up to 81%. Nevertheless, the MEC technology is comparatively a novel, emerging technology and thus presents some limitations. Production of hydrogen is a costly affair, when compared with production of other fuels. Moreover, its yield is also an issue, that needs to be resolved to provide a cost-effective solution (Nath and Das, 2004). There are few other limitations associated with biohydrogen production that needs to be addressed. Absence of commercial scale technology exhibition is one of them along with the further development and organization of storage options, transport system and utilization schemes for hydrogen (Balat and Kirtay, 2010).

8.4.2 Biobutanol

Butanol is a fuel of choice for transportation in the prevailing transportation sector driven by fossil-fuel. Recent decades have witnessed the popularity of biobutanol as a biofuel and is used in form of blend (butanol/gasoline blend) for transportation sector. The best part associated with the usage of fuel butanol is that its transportation can be easily managed in prevailing gasoline pipeline. Properties of butanol are eye catching and far better than that of ethanol. Butanol is an alcohol with four carbons. Its energy content is higher, its corrosivity is low and its volatility and hygroscopicity is less when compared with ethanol (Wu et al., 2013).

Conventionally, biobutanol was produced through ABE fermentation, in which carbohydrates were digested by Clostridium strains under anaerobic conditions to produce an amalgam of acetone, butanol, and ethanol. This method of producing butanol has certain limitations that makes it less preferred in comparison to the synthetic production of butanol from petrochemicals. Some of these challenges includes low yield, feedstocks compatibility, inhibition of products, slow rates of fermentations and soaring energy demanded for the product's down streaming (Kumar and Gayen, 2011). These limitations have forced the desire for evolving butanol through numerous novel paths excluding ABE fermentation, that would also support its commercialization.

These paths include the ethanol condensation catalytically, incorporation of genetically altered butanol-yielding microbial strains into new fermentation procedures, appliance of approaches including metabolic engineering, introducing continuous processes with simultaneous fermentation, and product recovery through methods like gas stripping, pervaporation, perstraction, adsorption. First generation feedstocks like sugarcane, sugar beet, and cereal crops can be employed for production of biobutanol. However, this would create food vs fuel controversies. To avoid these controversies, biobutanol can be produced potentially from second-generation feedstocks like lignocellulosic materials, which has unravelled a novel area for the advancements in cost-effective production of biobutanol. Barley straw, corn fiber, corn stover, wheat straw, switchgrass, degermed corn, domestic organic waste, sago, defibrated-sweet-potato-slurry (DSPS), extruded corn, cassava, liquefied corn starch and whey permeate are the examples of second-generation feedstocks that have been explored for biobutanol production via ABE fermentation (Kumar and Gayen, 2011).

8.4.3 Bioethanol

Biofuels have the potential to curtail the emission of greenhouse gas upto 85%, that is a significant percentage. Ethanol is one such option that can potentially alleviate the problems related with rise in global demands for energy. Ethanol is a trusted as an alternative fuel for a coming time and hence is produced in huge quantity (about 14-26 million tons) globally. There are many plus points associated with ethanol, when it is used as a fuel, chief among them is its high-octane value and its oxygenated nature. Moreover, the performance of ethanol is superior when compared to other fuels because it operates combustion engines at higher compression ratios. Ethanol is blended with petrol in different proportions and is named accordingly, for example E10, E85, E95 means, ethanol is blended with petrol in which it is 10%, 85% and 95%. When used in its pure state, it is designated as E100. When 10% ethanol is mixed with petrol it demands no alterations in engine.

Mixing of ethanol with gasoline, significantly curtails the demands for gasoline, which is non-renewable petroleum product. Ethanol is far a better fuel when compared to the gasoline, there are many reasons for that (Table 8.3)

Table 8.3. - Difference between ethanol and gasoline as a fuel (Chandel et al., 2017; Bhatia et al., 2018).

Characteristics	*Ethanol*	*Gasoline*	*Significance*
Flammability in air	1.3 to 7.6% v/v	3.5 to 19% v/v	Number and severity of vehicle fires are less for ethanol
Reid vapor pressure	16 KPa	71 KPa	Evaporative emission of ethanol is lower
Range of heating values	21.2 and 23.4 MJ/L,	30.1 and 34.9 MJ/L	Ethanol has a lower energy density
Need of adding Methyl tertiary butyl ether (MTBE)	Absent	Present	No contamination of ground water in case of ethanol

There are many microorganisms of industrial significance, that are potential producers of ethanol due to their capacity to metabolize various sugars. It covers broad spectrum of organisms belonging to yeast, bacteria and fungi. *Saccharomyces cerevisiae* is the most popular organism, known from ancient times for its capacity to generate ethanol. It is capable of metabolizing both glucose and sucrose. It is also normally identified as safe (GRAS). The only limitation associated with this organism is its incapability to metabolize pentose sugars. However, this incapability is of no meaning in the era of genetic engineering, where it is possible to engineer Saccharomyces with the genes involved in metabolism of pentose sugar (Bhatia and Johri, 2014).

Pichia stipitis (now known as Saffaromyces stipitis) is another yeast of industrial significance as far as ethanol production is concerned due to many plus points associated with this organism. This yeast is incapable of metabolizing polymeric cellulose as carbon source (Jeffries et al., 2007), but can potentially produce of many enzymes like cellulases (β-glucosidase and endoglucanases) that permits the yeast to ferment cellobiose and theoretically it can be presumed that recombinant strains are capable of hydrolyzing amorphous cellulose. S. stipitis also produces hemicellulases. These enzymes help in breaking down wood into simple (monomeric) sugars. This yeast is a well-known producer of xylanases and its xylanolytic ability was improved when the genes encoding for xylanase and xylosidase were co-expressed. This was prominently seen when employing birchwood glucuronoxylan was used as sole carbohydrate source in media. *Pichia stipitis* is popular for its potential to ferment xylose and can also act on variety of monomeric sugars prevalent in lignocellulosic biomass. There are various toxic substances, that are generated during pretreatment of substrates like acetic acid, furfural and hydroxymethylfurfural (HMF), that have an adverse effect on the physiology of various fermenting organisms. P. stipitis is capable of utilizing acetic acid and reducing the furan ring found in furfural and HMF, thereby creating a chance for this organism to clean up these lethal substances during cellulosic biomass conversion (Bhatia et al 2019). *P. stipitis* has been explored by various researchers to produce ethanol from various substrates under different parameters (Table 8.4)

Table 8.4 - Potential of *P. stipitis* strains to produce ethanol from various substrates (Agbogbo and Coward-Kelly, 2008)

Substrates	*Parameters*	*Strain of P. stipitis used*	*Ethanol Production*
Prosopis juliflora	hemicellulosic hydrolysates (18.24 g sugar/L broth)	P. stipitis 3498	7.13 g/L ethanol.
L. camara	pH 5 and 30°C for 36 h	P. stipitis 3498	0.33 g alcohol/g lignocellulose used
Water hyacinth	detoxified hemicellulose acid hydrolysate (rich in pentose sugars), pH 6.0 and 30°C	P. stipitis NCIM-3497	0.425 g ethanol/g lignocellulose

Along with *Pichia stipitis,* there are many other pentose fermenting organisms like *Candida shehatae* and *Pachysolen tannophilus* that are promising for industrial applications. Yeasts are able to metabolize xylose with the aid of the xylose reductase (XR) that transforms xylose to xylitol. Later on, xylitol dehydrogenase (XDH) transforms xylitol to xylulose. Post phosphorylation, xylulose gets utilized via the pentose phosphate pathway (PPP). P. stipitis is capable of performing xylose fermentation under anaerobic conditions because it has xylose reductase capable of using both the cofactors i.e., NADPH and NADH. NADH is used under anaerobic conditions (Bhatia and Johri, 2015a). Sanchez et al., (1999) reported the use of *Pachysolen tannophilus.* Attempts were made to develop a yeast that was capable of producing industrial ethanol by xylose fermentation. Induction of UV light, generated the better strains of *P. tannophilus* that could faster on xylose. Numerous other yeast strains capable of utilizing xylose have been reported (Bhatia and Johri, 2018).

Zymomonas mobilis is also a known potential producer of ethanol on commercial scale due to many of its physiological merits. It has simple nutritional requirements and is able to uptake glucose, fructose or sucrose efficiently. However, it prefers glucose to other sugars. This organism has gained GRAS status and is considered superior to S. cerevisiae due to its capacity to metabolize wide range of sugars (Bhatia, 2019).

Various strains of *Kluyveromyces marxianus* have potential to grow at high temperature and ferment cocktail of sugars such as glucose, xylose, mannose and galactose to produce ethanol (Fonseca et al. 2008). For instance, thermotolerant strain *Kluyveromyces marxianus* 6556 can perform Simultaneous Saccharification and Fermentation (SSF) of lignocellulosic agroresidue at 37°C. Strains of *K.*

marxianus can grow at elevated temperatures of 52°C where they exhibit a high growth rate and short generation time. At high temperature, *K. marxianus* can potentially transform a broad spectrum of substrates, counting xylose, to ethanol by performing efficacious SSF with a diverse feedstock. This organism possesses genes that encodes thermotolerant enzymes like endoglucanase, cellobiohydrolase, and β-glucosidase, as a result of which, strains are able to thrive well in synthetic media comprising cellobiose or carboxymethylcellulose as solitary source of carbon. But the breakdown of crystalline cellulose was not observed. Recent studies indicate that when *K. marxianus* strain could effectively transform 10 g/l of a cellulosic β-glucan to 4.24 g/l ethanol at 48°C in span of 12 h when it was genetically altered to display endoglucanase II of *Trichoderma reesei* and β-glucosidase of *Aspergillus aculeatus* on its cell surface (Bhatia et al., 2012). Bhatia and Johri (2015b) have demonstrated *Mucor indicus* as a potent producer of ethanol.

There are many plus points associated with microalgae, that makes them a prospectus cell factory to biosynthesize bioethanol. They possess thin cellulosic cellular wall wide range of metabolic diversity, prominent photosynthetic potential, immense biomass productivity, varying chemical cellular constituents incorporating higher sequestration of carbon dioxide. Two metabolic pathways are prominent in microalgae that leads to generation of bioethanol. First is fermentation and second is gasification. Cell wall of microalgae is rich in starch and thus acts as an efficient feedstock for bioethanol production in bioethanol generation in alliance with yeast *Saccharomycess cerevisiae.* Acid, alkali or enzymatic treatment of microalgal biomass, speeds up the release of simple residual sugars, which thereafter gets biotransformed to ethanol via yeast fermentation. Species of *Chlorococcum, Chlorella, Chlamydomonas, Scenedesmus, Tetraselmis* have significant amount of carbohydrates which are valuable for producing bioethanol comprehending above-mentioned conceptualization (Rodrigues et al., 2015). *Chlorococcum humicola* is a potential microalga for ethanol fermentation. *Chlorococcum humicola* generates $7.2 g.L^{-1}$ ethanol in view of a mixed consortium method in alliance with *S. cerevisiae* (Eduardo and Alberto, 2016). In parallel fashion, *Chlorella vulgaris* microalgal species is considered as a one of the highest ethanol producing microalgae ($238 dm^3.mg^{-1}$) (Doan et al., 2012). *Scenedesmus dimorphus* comprise 53.7 w/w carbohydrates with the hydrolysis of 80% fermentable sugar, and is considered as a potent stock of bioethanol (Khan et al., 2018).

8.4.4 Biomethane

Biomethane is a biofuel, also known by the name green gas and synthetic natural gas. Hydrolysis of organic material under anaerobiosis leads to its production. Three important steps are concerned in methane production under anaerobic condition. The first step involves the acidogenic microbes, that secretes the hydrolytic enzymes and causes the hydrolysis/liquefaction reactions in this step. In this step insoluble complex organic molecules are hydrolysed into soluble monomeric substances like monosaccharides, fatty acids, and amino acids. Cellulose and lignocellulose structures gets opened during optimum pre-treatment (Fig. 8.3). Second step is acidogenesis, that involves acidogenic microorganisms that converts the hydrolysed products / monomers into simple compounds organic acids, CO_2 and H_2. The chief acids produced during acidogenesis are propionic acid, acetic acid, butyric acid and ethanol. The third and last step is methanogenesis stage where methane is produced either by two ways. The first way is breakdown of acetic acid to CO_2 and methane (CH_4), and the second way is the reduction of CO_2 with hydrogen (Molino et al., 2013).

The major component of generated biogas is methane (50–75%) followed by CO_2 (25–50%). The remaining is constituted of water vapour, and traces of O_2, N_2 and H_2S. Biomethane is separated from rest of the components at the stage of purification, in which rest of the impurities like H_2S, NH_3, CO_2 and siloxane are removed. Purification helps in improving the concentration of methane, that reaches upto 96% in biomethane, which matches the quality standards of natural gas.

All sorts of biomass comprising cellulose, hemicellulose carbohydrates, proteins and fats are suitable feedstock for biogas production. The factors that influence the methane yield and biogas components are feedstock nature, digestion system and retention time (Ranalli, 2007). As mentioned earlier, lignocellulosic biomass (LB) is comprised of lignin, cellulose and hemicellulose. Few studies have revealed that the LB that has high percentage of lignin, have a methane potential (0.18–0.32 m^3 methane/kg volatile solids) (Lehtomäki et al., 2006). Methane potential is also governed by solid content in the feedstock and crop yield. Besides lignocellulosic crops, agro-residues (like bagasse) and agricultural feed stocks are also a potential cost-effective lignocellulosic feedstock. Other potential feedstocks are forest feedstock like softwood and hardwood generated out of wood processing plants (Sims et al., 2008). However, a significant cost is involved in pre-treatment of the LB to produce fermentable sugars that creates a quandary with these feedstocks. This cost is substantially more than that involved in production with first generation biofuels. Variety of pre-treatment strategies helps in hydrolysing or liquefying the substrate, thereby increasing the methane yield potential. Some of these pre-treatment methods include steam, lime, liquid hot water and ammonia that promotes the production of methane from LB (Hendriks and Zeeman, 2009). The further evaluation of these pre-treatment methods is needed to establish energy cost involved in pre-treatment versus net energy gain to yield methane. It is also important to consider the cost involved with purification step of converting biomass to biomethane, its storage and its transportation at end point of its usage. This analysis will determine the overall cost benefit of using LB.

To enhance the proportion of methane in final product, raw biogas demands purification to remove various impurities prevalent during production (Table 8.5). As mentioned earlier, CO_2 shares the significant proportion of the biogas that declines its calorific value and escalates the cost involved in its compression and transport. Conditioned production of biomethane is now feasible due to the availability of commercial membranes for onsite purification. Cost and economics are the chief criteria for the commercial execution and implementation of any technology.

Table 8.5 - Current Methods of purifying the impurities associated with biomethane in raw biogas (Ryckebosch et al., 2011; Niesner et al., 2013).

	Impurities associated with biomethane in raw biogas			
	Hydrogen sulfide	**Siloxanes**	**Ammonia**	**Carbon dioxide**
Current Methods of their purification	chemical absorption	absorption	washing with dilute HNO_3 or H_2SO_4	water scrubbing
	aeration, membrane separation	adsorption on activated carbon	wet chemical scrubbing	pressure swing adsorption
	biological filters	cryogenic separation	activated carbon filtration	chemical scrubbing
	adsorption on activated carbon	membrane separation		physical scrubbing
	chemical oxidation			

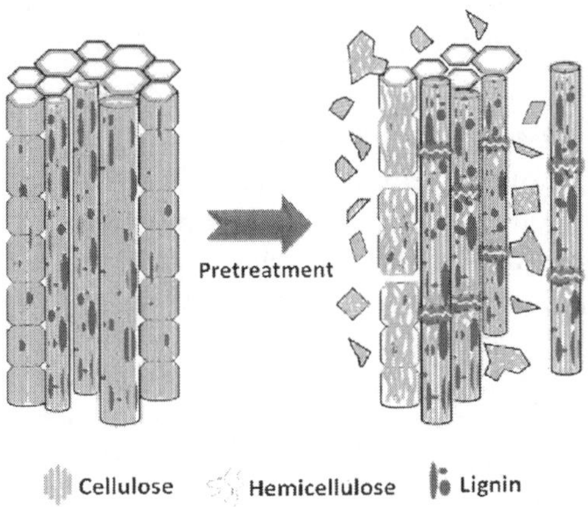

Figure 8.3. - The main aim of pretreatment

8.4.5. Biodiesel

Biodiesel is also an important biofuel, and presently microalgae are employed as potential biocatalysts for its production. These organisms are able to generate biodiesel by metabolizing broad range of waste matter under diverse environmental accessibility zone. Nevertheless, expenditure involved in downstream processing is huge that raises the production cost of biodiesel production from this organism. This is the major limitation. Microalgae not only contribute towards biodiesel production but also curtails the atmospheric CO_2 level for reducing the impression of air pollution and global warming. Numerous steps are concerned during the production of biodiesel from microalgae, viz. transesterification, lipid rich biomass production, harvesting, microalgal cell disruption, extraction, drying from oil bodies, etc.) (Katiyar et al., 2017). Other than microalgae, there are several feedstocks that are potentially employed for biodiesel production, like soybean oil, palm oil, rapseed oils etc that are used as initial substrate.

The major limitation associated with the usage these substrates (i.e., soybean oil, palm oil, rapseed oils) is that they cause interreference with the food chain in the human community. However, no such competition exists, when microalgae are employed for biodiesel production. Microalgae-based transesterification reactions are involved in the conversion of lipids/ triacylglycerols / free fatty acids into biodiesel. Crude oil of microalgae is viscous. The transesterification process helps in conversion of high molecular weight free fatty acids into low molecular weight. In the biological process of transesterification, methyl alcohol (C1 alcohol) reacts with crude algal oils in presence of catalyst to obtain fatty acid methyl ester (FAME) precursor. Throughout this process, acid-alkali both is employed as catalyst but some researches have reported that alkali is unsuitable for microalgal lipid transesterification. FAME then gets finally converted to biodiesel along with glycerol as chief by product. Different methods are employed for extraction such as physical extraction, chemical extraction and extraction of fatty acid-based bio-oils. These methods govern the generation of fatty acids, biodiesel and algal cellular mass (Bux, 2013).

Scenedesmus acuminatus has been contemplated as a cost effective microbial cellular factory for producing biodiesel. *Scenedesmus bijuga* can utilize various arrays of waste waterbodies for producing biodiesel. Consortia of *Phaedoactylum tricornutum* and *Porphyridium purpuruem* displays a great possibility towards biodiesel synthesis. Lipid generation rates of *Nannochloropsis gaditana, N. granulate, N. salina, N. ocenaica, N. oculata* are excellent (158.76 ± 13.83 mg.L^{-1}.d^{-1}), thus making them prospectus candidate for biodiesel production (Bhalamurugan et al., 2018). *Chlorella vulgaris* is comprised of rich quantity of starch (~37% dry wt), saturated fatty acids (22.65 %), unsaturated fatty acids (~77%), C14-C18 chain length of fatty acids, and ethanol (65 %), that makes it an ideal candidate biodiesel production. This alga produces 3726 kg-water/kg biodiesel in course of the open pond systems (Maeda et al., 2018). Some researchers have revealed that close system photobioreactor is a better producer of microalgal biomass in comparison to open system photobioreactior. Latter is found more suitable for producing biodiesel *Scenesdmus* sp. are also a potent producer of biodiesel, as the surface level in these species have lipid layer, alcoholic groups, carboxyl groups which are proficient for producing biodiesel. Similarly, species of Euglena, Spirogyra, Scenedesmus, Pseudokirchneriella, Phormidium, Nitzschia and Desmodesmus contain huge amount of unsaturated fatty acid and fatty alcohols that are major backbone for producing biodiesel (Sudhakar and Premalatha, 2015).

8.5 Platform Chemicals

Various global agencies and national government entities are in support of a conversion from present fossil-based economy towards a bio-economy which is sustainable. The organisation that emphasis on this shift are Organization for Economic Co-Operation and Development (OECD), International Energy Agency (IEA), World Economic Forum (WEF) and National Renewable Energy Laboratory (NREL). These organisations support this transition, as several benefits are associated bio-economy (Sudhakar et al., 2018). Bioeconomy is an cost effective, eco-friendly and a sustainable approach. It is eco-friendly as it helps in curtailing the emanation of greenhouse gases and lessen the dependencies on non-renewable fossil fuel resources (de Jong et al., 2013). Bioeconomy supports the development of socially, environmentally and economically a more sustainable global economy. The expansion of robust biorefinery systems comprising innovative technologies is needed for the extension of an efficacious bio-economy. Advanced biorefineries can successfully process biological feedstocks like lignocellulosic substrates into a diversity of bio-based products like biofuels and higher value-added products in an effective and cost-effective manner (Nanda et al., 2015). New business opportunities are created and driven when advanced biorefineries are developed at industrial scale (Pandey, 2011).

When a chemical act as a substrate for the manufacture of diversity of other higher value-added products, it is known as a platform chemical. In the year 2004, twelve chemical building blocks were identified as potential platform chemicals by the Department of Energy (DOE). Biomass was the source of these chemicals (Werpy et al., 2004). The list of platform chemical was updated by DOE in 2010 that comprise glycerol, ethanol, lactic acid, furfural, hydroxymethylfurfural, sorbitol, isoprene, 2,5-furandicarboxylic acid, xylitol, succinic acid, levulinic acid and 3-hydroxypropionic acid/aldehyde (Bozell and Petersen, 2010). Except glycerol and isoprene, all these recognized platform chemicals can be generated from carbohydrate sources derived from biomass. According to McKinsey & Co. guesstimates, the bio-based product sale in 2012 were $252 billion and that of renewable-based chemical was nearly 9% of global chemical sales and are anticipated to grow by 4% yearly. The sales of bio-based product are anticipated to escalate with an yearly growth rate of 8% to $375 to $441 billion by 2020 (BIO, 2016).

8.5.1 Glycerol

Glycerol is also known as glycerin. It is colourless and odourless organic liquid that is of immense use in industries such as cosmetic, food, pharmaceuticals etc. There are many processes that produces bioglycerol, such as biodiesel production, soap production and microbial fermentation. Glycerol has been categorized into two groups, based on purity. Crude glycerol is 60-80% pure and refined glycerol is 99.1-99.8% pure. Commercially important glycerol needs to be extremely pure, particularly pharmaceutical industry can only use glycerol, if it is more than 99%

pure. Impurities like methanol associated with glycerol needs to be removed when glycerol is needed for pharmaceutical industry. Removal of methanol and other impurities make glycerol ultrapure. Glycerol purification is a cost intensive process and many a times it is not feasible for small firms to go for it. Many refinery processes can be used to purify glycerol. Some of them are decantation, chemical treatment, distillation, ion exchange, adsorption, extraction and crystallization. There is a noteworthy difference in the global price of pure and crude glycerol. Price of former varies from \$1.10/kg to \$3.30/kg, whereas the latter varies from S0.04/kg to \$0.33/kg. As there had been a global boom in biodiesel production in past decades, there had been a simultaneous boom in the production of crude glycerol and pure glycerol (Tabah et al., 2016).

Nearly 10% (w/w) of glycerol is produced in course of biodiesel production process; leading to about 90% of over-all glycerol produced (Okoye and Hameed, 2016). Due to its easy accessibility in huge quantity and its prospective nature to act as a primary building block in a biorefinery glycerol has gathered significant consideration in its alteration to the higher value-added chemicals. One such alteration is the catalytic hydrogenation of glycerol in the existence of Ruthenium/ Carbon or Platinum/Carbon to deliver ethylene propylene glycol, glycol and acetol (Nakagawa et al., 2014). The companies that are producing renewable propylene glycol from glycerin are Cargill, Dow Chemical Company, Archer Daniels Midland and Huntsman. Another example of transforming glycerol to a value-added product is its reduction in presence of -OH bases to produce lactic acid as the primary product. Various research groups have also explored the biochemical transformations of glycerol to produce 1,3-propanediol (1,3-PDO). Fermentation of glycerol with genetically engineered Clostridium acetobutylicum offers a significant yield of 1,3PDO.

Fermentation of glycerol with Escherichia coli produces 1,2-propanediol. Applications of propanediols is depicted in figure 8.4. glycerol carbonate is another important derivate of glycerol that gets produced when glycerol reacts with ethylene / propylene carbonate / urea or carbon dioxide (Okoye, 2016). Glycerol carbonate has a broad spectrum of relevance in the production of industrially significant chemicals like glycidol, and in polymers, adhesives, coatings and lubricants (Fig. 8.5). Epichlorohydrin is also commercially produced from glycerol. Two equivalents of HCl reacts with glycerol in the existence of CH_3COOH (acetic acid) to primarily produce 1,3-dichloro-2-propanol (Santacesaria, et al. 2010). Epichlorohydrin is manufactured by Solvay and Dow Chemical Company employing an alike process. Plus points associated with this process that makes it superior in comparison to traditional manufacturing protocols are the production of less chlorinated waste and usages of 90% less water overall. Filamentous fungi and yeast fermentation helps in the conversion of glycerol into several organic acids and polyols. Clomburg and Gonzalez (2013) have reviewed the anaerobic fermentation of glycerol to numerous acids and alcohols. Likewise, 2,3-butanediol, 1-butanol, ethanol, lactic acid, 1,3-propanediol, propionic acid, dihydroxyacetone and succinic acid can be synthesized by exploiting variety of microorganisms (Abad and Turon, 2012). The compounds of commercial significance like glyceraldehyde, glyceric acid, and tartronic acid are obtained when selective oxidation of the primary hydroxyl groups of glycerol takes place. Whereas, dihydroxyacetone (DHA) is produced when secondary hydroxyl group gets oxidized and ketomalonic acid is produced when all three hydroxyl groups get oxidized (Gil et al., 2014). Another derivative of glycerol is glycidol. It is a substance of industrial engrossment, as it can potentially produce chemicals of industrial significance like epoxy resins, polyurethanes, and polyglycerol esters.

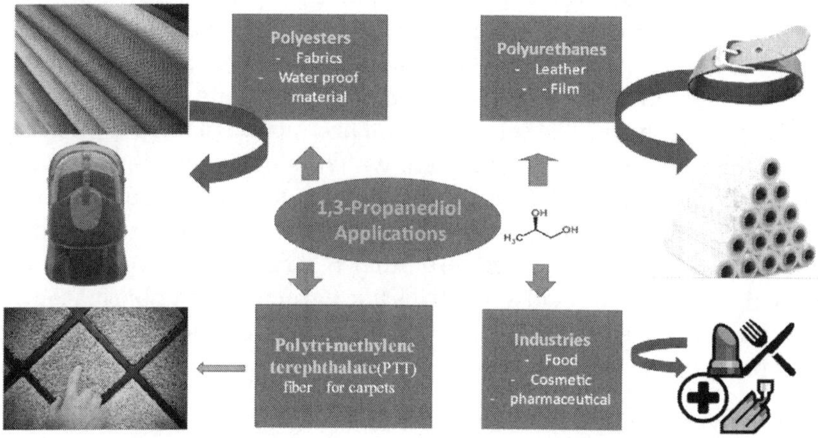

Figure 8.4. – Potential uses of 1,2-propanediol in various Industries

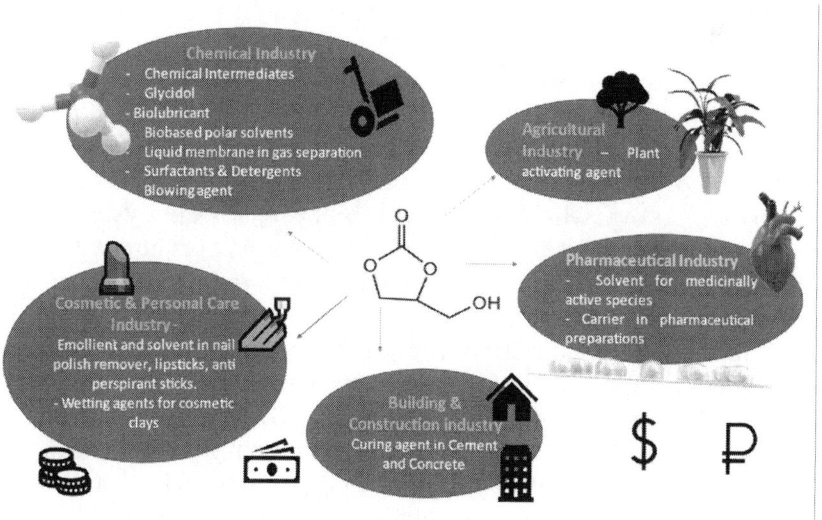

Figure 8.5. – Potential usages of glycerol carbonate in numerous industries

8.5.2 Furfural

Furfural is formed in a lignocellulosic biorefinery from numerous renewable agro sources. Global production of this platform chemical is assumed to be around 300,000 tonns annually. Commercial production of bio-based furfural occurs through acid mediated dehydration of xylose. In the first step, C5 sugars (especially xylose) gets generated when hydrolysis of C5 polysaccharides found in hemicellulose of lignocellulosic biomass takes place in presence of mineral acids. This xylose then gets transformed into furfural due to its dehydration, in presence of acid (Verma et al., 2017). Various lignocellulosic biomasses like sugarcane bagasse, corn stalk and eucalyptus wood have been reported to produce furfural in presence of different concentrations of mineral acids. Solvents have been tested both in their monophasic system and biphasic system. Examples of former are supercritical fluids, organic solvents and ionic liquids, and the example of latter is mixture of water and organic solvent. The selectivity and production of furfural was improved in presence of 2-methyltetrahydrofuran, which is a renewable organic solvent tested in a biphasic system. Majority of share of furfural production (about 70%) is employed as a chemical feedstock for the furfuryl alcohol production. Additionally, furfural can similarly be converted into other 5-membered oxygen heterocycles, succinic acid, and levulinic acid. Industries like agrochemical, pharmaceutical and plastics extensively use furfural. Moreover, furfural is also employed as adhesives, and flavor enhancers (Mariscal et al., 2016).

8.5.3 Hydroxymethylfurfural (HMF)

HMF can be formed from broad array of substrates like monosaccharides, polysaccharides and pretreated biomass in presence of numerous acid catalysts. High yields of HMF gets produced when fructose is dehydrated in existence of acid catalyst. Among the several organic solvents tested for producing HMF, dimethyl sulfoxide (DMSO) was found utmost suited for converting sugars into HMF (Rout et al., 2016). The only limitation with the use of DMSO is its high melting point that poses a challenge to separate the product. When water was tested as a solvent or reaction medium, in existence of various acid catalysts, the product produced was about 50%. This problem was because of product degradation in water monophasic solvent system. To overwhelm this effect, biphasic solvent systems have been investigated, in which there is a continuous extraction of product into the organic layer, thereby evading any degradation of the anticipated HMF product. Moreover, the biphasic solvent system aids in improving selectivity and yield of HMF in the course of facilitating its separation from the reaction medium. Numerous value-added polyester building blocks can be prepared from HMF. Chief among them are 2,5-dimethylfuran, 2,5-furandicarboxylic acid (FDCA), 5-ethoxymethylfurfural, ethyl levulinate, γ-valerolactone and 2,5-bis(hydroxymethyl)furan (Li et al., 2016). 1,6-hexanediol (1,6-HDO) can also be prepared from HMF and is employed in preparing polycarbonatediols. Polycarbonatediols are used in producing polyurethanes for application in elastomers, coatings and adhesives. Furthermore,

1,6-heaxanediamine and ε-caprolactone can also be prepared from 1,6-HDO which are employed in the production of numerous polymers (Isikgor and Becer, 2015).

8.5.4. 2,5-Furandicarboxylic Acid (FDCA)

FDCA is an excellent alternative for petrochemical-founded terephthalic acid and adipic acid, and thus has been investigated significantly. HMF oxidation, heads to the production of FDCA employing numerous homogeneous and heterogeneous systems (Chatterjee et al., 2015). If fructose, is the starting material, FDCA is synthesized in two-step dehydration. In first step dehydration of fructose takes place to generate HMF as an intermediate. Further dehydration of HMF (without isolating it) leads to formation of FDCA. FDCA has numerous probable relevance in polyesters, polyamides, and plasticizers. The chemical, physical and mechanical properties of polyethylene furanoate (PEF) polymers produced from FDCA and ethylene glycol are parallel to the petroleum-based polymers terephthalate and polyethylene (Isikgor and Becer, 2015). Pilot plant for generating FDCA and polyethylene furanoate polymers are operated by Avantium. A commercial production plant is built with the joint venture of Avantium with BASF, that has a yearly capability of up to 50,000 metric tons/ year (Avantium, 2016). Archer Daniels Midland (ADM) and Dupont have established a high generating procedure for manufacture of FDCA methyl ester straight from fructose (DuPont press release, 2016).

8.5.5 Succinic Acid

Traditionally, production of succinic acid was accomplished through the hydrogenation of petroleum-based maleic acid or through butanediol oxidation. Recent years have witnessed the commercial production of succinic acid through the biochemical fermentation of biorefinery sugars by some companies. Fermentation of glucose by E. coli, is the base for commercial scale production of succinic acids by firms like Bioamber (press release 2015) and Roquette/DSM (press release 2008). In the year 2013, market price of succinic acid was nearly $115 million and it is anticipated to touch $1.1 billion by the year 2021. Succinic acid is a platform chemical can be transformed into succinate esters, which are base material for production of tetrahydrofuran, 1,4-butanediol and γ-butyrolactone (Delhomme et al., 2009).

Succinic anhydride is a product of dehydrogenative cyclization of succinic acid. Succinic anhydride is a chief base material for producing maleic acid and fumaric acid (Delhomme et al. 2009). An alliance was established between NatureWorks and Bioamber to investigate the production of entirely bio-renewable polyester copolymers of 1,4-butanediol and succinic acid (Adkins et al., 2012). Poly (propylene succinate) (PPS), poly (ethylene succinate) (PES) and poly (butylene succinate) (PBS) were among the various polyesters that can be produced, that were extensively studied and were efficaciously commercialized (Isikgor and Becer, 2015).

8.5.6 Lactic Acid

Lactic acid (LA) is industrially manufactured through the fermentation of numerous sugars like sucrose, glucose or lactose (Ghaffar et al., 2014). Corbion produces l-lactic acid by fermentation of carbohydrates (Corbion). Calcium lactate (90% yield) is produced in the course of fermentation process. Neutralization of calcium lactate furnish lactic acid. Calcium sulphate is produced during neutralization process at the rate of one equivalent, which poses a problem for disposal. Recent technologies such as ion-exchange resins and nanofiltration have been evaluated to overcome the problems associated with neutralization.

Current years have observed the development of processes in which xylose can be converted into lactate employing engineered yeast. Additionally, pilot scale production of lactic acid was done by Direvo Industrial Biotechnology, in which single step employing a consolidated bioprocessing technology produced lactic acid directly from lignocellulosic biomass (Direvo press release, 2013). The global production of lactic acid is around 350,000 tons/ year and is assumed to escalate significantly in the subsequent decade. Performance properties of polylactic acid polymer resembles that of polystyrene or polyethyleneterephthalate, in terms of its heat stability and outstanding barrier for flavors. Polylactic acid is popular among the various biodegradable polymers in terms of its accessibility, cost effectiveness and potentiality. Production of lactic acid on commercial basis is presently done by many companies, with major portion of its production being utilized by textiles and packaging market (Isikgor and Becer, 2015).

8.5.7 Levulinic acid

Levulinic acid is generated in significant quantity during hydrolysis of C6 sugars in presence of acid. Recent years have witnessed the commercial scale production of levulinic acid out of lignocellulosic feedstock (Pileidis and Titirici, 2016). Different companies are producing levulinic acid employing different strategies. For instance, Maine Bioproducts produces levulinic acid by two stage process in which dehydration of lignocellulosic feedstock takes place in presence of acid. GF Biochemicals employing their proprietary technology, initiated the industrial production of levulinic acid from lignocellulosic biomass in 2015 (GF Biochemicals press release, 2015). Avantium synthesised levulinic acid methyl ester by developing a process that starts from plant carbohydrates (Avantium). There are many other companies that have showed their interest in producing levulinic acid-based biopolymers. For instance, Segetis (acquired by GF Biochemicals press release 2015) explored the usage of ketals made from levulinic acid in thermoplastic and polyurethane applications (Rose and Palkovits, 2011). Applications like pharmaceuticals, plasticizers, fragrances, and cosmetics use levulinic acid as a building block. Levulinic acid has the potential to substitute petro-based chemicals. For instance, diphenolic acid (DPA) derived from levulinic acid can behave as a alternative for bisphenol-A (BPA) in consumer products and food containers. Moreover, higher value-added products can be formed from levulinic acid, for example valeric acid, levulinic acid esters, γ-valerolactone, 5-aminolevulinic acid and 2-methyltetrahydrofuran (Pileidis and Titirici, 2016).

8.5.8 Xylitol

Xylitol is a natural 5-C sugar alcohol, that is 20% more sweeter than sucrose, whereas its calories output is just 40%. Moreover, metabolism of xylitol is insulin independent, and thus it is considered safe substitute for sugars for diabetic people (Lugani et al., 2017). Xylitol is industrially produced through the catalytic reduction of pure d-xylose derived from hemicellulosic hydrolysate of hardwood in the existence of Nickle catalyst at elevated temperature and pressure. Although commercial production of xylitol is mediated through chemical reductive methods, numerous reports have described its production through biochemical reductive methods employing either microorganisms or enzyme. Process of xylitol production has become compacted in recent times, where only one vessel method is involved to convert hemicellulose into xylitol utilizing an acid in presence of ruthenium (Ru) on carbon catalyst (Dietrich et al., 2017). Correspondingly, the alteration of corncob-derived hemicellulose into xylitol is achieved employing a Ru catalyst maintained on carbon nanotubes (Ribeiro et al., 2016).

8.5.9 Isoprene

Isoprene is a 5-carbon unit for the production of a diversity of polymers. The most of the isoprene synthesized is transformed into the polyisoprene polymer which is then employed in diverse products like medical appliances, footwear, sporting goods, mechanical instruments and rubber tires. Though the industrial manufacture of isoprene is presently supported from petroleum-based feedstocks, production procedures based on renewable resources are in progress. A combined approach to produce bio-isoprene from renewable raw materials on a commercial scale (Genencor) is being codeveloped by the Goodyear Tire Company and Dupont–Genencor. Their technology employs a genetically modified E. coli for the fermentation process (Whited et al., 2010). Plants sugars are employed for producing isoprene, the development and commercialization of which is explored by Amyris in corporation with Michelin and Braskem (Amyris).

8.5.10 3-Hydroxypropionic acid (3-HP) and 3-hydroxypropionaldehyde (3-HPA)

3-HP and 3-HPA are platform chemicals of industrial significance as they can be converted into value-added products or chemicals. Though many researches are in progression, presently there is no processes of production of 3-HP and 3-HPA from renewable sources. Toxicity of 3-HPA poses a challenge for its biological production. Presently, fermentation of glycerol produces 3-HPA, that upon oxidation yield 3-HP. 3-Hydroxypropionic acid is directly generated from glycerol by employing the strains of Klebsiella pneumoniae and E. coli. E. coli can ferment glucose to yield 3-HP at neutral pH, whereas yeast S. cerevisiae does the same at acidic pH (Kumar et al., 2013). Production of 3-HP (pKa of 4.51) at low pH is advantageous, as this process not only improve the economics, but the amount of waste generated is also significantly low (Kildegaard et al., 2015), thus the process based on yeast is desirable. As a platform, chemical 3-HP and 3-HPA can be transformed into numerous value-added chemicals like acrylic acid, acrylic acid esters and amides, acrolein, malonic acid, 3-hydroxypropionic esters and 1,3-propanediol. Various polymers are synthesized from acrylic monomer. Companies like Dow Chemical Company, BASF, Lubrizol and AkzoNobel commercially produce Polyacrylic acid. The treatment of 3-hydroxypropionaldehyde with an amine yield acrylamides. Monomer of acrylamide is lethal, whereas its polymer polyacrylamide is non-lethal and finds its applications in treatment of water, manufacture of paper, oil recovery, mining, absorbents and as a gel used in electrophoresis. Acrolein and acrylonitrile are two derivatives of industrial significance which are employed for synthesizing various polymers (Zheng et al., 2006).

8.5.11 Sorbitol

Sorbitol is the most normally consumed sugar alternate with an estimated yearly global production of 800,000 tons. Sorbitol is extensively expended in industries like food, drugs, beverages, cosmetics, and to yield value-added chemicals (Zhang et al., 2013). Commercial production of sorbitol involves hydrogenation of d-glucose catalyzed by transition metal (Ortiz et al., 2013). The commercial production of sorbitol involves 2 stages. In first stage, the raw material like wheat, cassava and corn undergoes enzymatic hydrolysis to yield glucose. In the second stage, hydrogenation of glucose takes place to yield the anticipated sorbitol. Though Nickle has been used as a catalyst in the previous industrial processes, few reports have stated the use of Ru, an outstanding catalyst in converting glucose into sorbitol with extreme selectivity (Zhang et al., 2013). Presently, Cargill, Roquette Freres and SPI Polyols are the chief producers of sorbitol. The share of Roquette Freres in this production is around 70% of the market volume. Recently, cellulose and cellobiose have also been explored to produce sorbitol by hydrolysis pursued by hydrogenation. In this conversion process, catalysts comprising an acid or heteropolyacid with sustained metal catalysts such as Ruthenium or Platinum, are used (Zhang et al., 2016). Nearly 15% of the sorbitol generated is exploited in the industrial translation into ascorbic acid by fermentation. Sorbitol dehydration in the existence of a Cu catalyst results in the production of isosorbide. Sorbitol on its hydrogenolysis with multifunctional catalysts yields alcohols of lower carbon number like methanol, ethanol, glycerol, propyleneglycol and ethylene glycol. These alcohols can then be further transformed into higher value-added products (Isikgor and Becer, 2015). Polymers can also be synthesized from sorbitol that have an extensive applications like biodegradable polymers and in biomedicines and biocomposites. Correspondingly, poly (isosorbide carbonate) and the isosorbide polymer, are believed as an assuring substitute to the petro-based BPA polycarbonate because to its exceptional characteristics (Roquette). Polymer grade isosorbide and isosorbide diesters are also produced by Roquette that are used as phthalate-free plasticizers (Roquette).

8.6. Conclusion

It can be concluded that the industrial production of fuels and chemical have taken new dimension and novel approaches are streamlined for their production. It is now a late concept to rely upon non-renewable sources for the production of chemicals and as an energy source. The concept of biorefineries have brought a platform for various biomasses particularly lignocellulosic biomasses to be explored for production of fuels and chemicals. Numerous magnanimous chemical companies such as BASF, DuPont and the Dow Chemical Company are intensely valorizing these lignocellulosic biomasses. Platform chemicals are now extensively produced and marketed on commercial basis as they have been the source of various value-added products. The need of an hour is to develop and extend the concept of integrated biorefinery which is a sustainable biorefinery producing bio-energies and high value added bioproducts. The manufacture of bio-based fuels and chemicals in a biorefinery can substantially curtail the emission of the greenhouse gas and is a sustainable approach towards cost effective, eco-friendly bioeconomy. Subsequently, it can be concluded that many chemical industries have contacted to innovative modulation point one in which renewable resource-constructed industries have exclusively complemented petro-chemical industries, and it is definite that bio-based methodologies will have a position in persuasion of economy.

References

Abad, S. & Turon, X. (2012). Valorization of biodiesel derived glycerol as a carbon source to obtain added-value metabolites: focus on polyunsaturated fatty acids. Biotechnol Adv, 30, 733–741.

Adkins, J., Pugh, S., McKenna, R. & Nielsen, D.R. (2012). Engineering microbial chemical factories to produce renewable "biomonomers". Front Microbiol, 3, 313.

Agbogbo, F.K., Coward-Kelly, G. (2008). Cellulosic ethanol production using the naturally occurring xylose-fermenting yeast, *Pichia stipitis*. Biotechnology Letters, 30, 1515-1524.

Alonso, D.M., Bond, J.Q. & Dumesic, J.A. (2010). Catalytic conversion of biomass to biofuels. Green Chem, 12, 1493–1513.

Avantium (2016). Press release. https ://www.avant ium.com/press -releases/ synvina-joint -venture-basf-avantium-established.

Balat, H. & Kirtay, E. (2010). Hydrogen from biomass—present scenario and future prospects. Int J Hydrogen Energy, 35, 7416–7426.

Beer, L.L., Boyd, E. S., Peters, J. W. & Posewitz, M.C. (2009). Engineering algae for biohydrogen and biofuel production. Curr Opin Biotechnol, 20, 264–271.

Bhalamurugan, G.L., Valerie, O. & Mark, L. (2018). Valuable bioproducts obtained from microalgal biomass and their commercial applications: A review. Environ. Eng. Res., 23, 229-241.

Bhatia, L. & Johri, S. (2014). Fourier transform infrared mapping of peels of Citrus sinensis var mosambi after physiochemical pretreatment and its SSF for ethanol production by Saccharomyces cerevisiae MTCC 3821- An economic & ecological venture. International Journal of Science & Research, 3 (11), 1653-1664.

Bhatia, L. & Johri, S. (2015a). Biovalorization potential of peels of Ananas cosmosus (L) Merr. For ethanol production by Pichia stipitis NCIM 3498 & Pachysolen tannophilus MTCC 1077. Indian Journal of Experimental Biology, 53, 819-827.

Bhatia, L. & Johri, S. (2015b). FTIR analysis & optimization of simultaneous saccharification and fermentation parameters for sustainable production of ethanol from peels of Ananas cosmosus by Mucor indicus MTCC 4349. Waste and biomass valorization. 7 (3), 427-438. doi: 10.1007/s12649-015-9462-4.

Bhatia, L. & Johri, S. (2016). Optimization of simultaneous saccharification and fermentation parameters for sustainable production of ethanol from sugarcane bagasse by Pachysolen tannophilus MTCC 1077. Sugar Tech. 18 (5), 457–467. doi: 10.1007/s12355-015-0418-6.

Bhatia, L. & Johri, S. (2017). Fourier transform infrared mapping of peels of Litchi chinensis after acid treatment and its SSF for ethanol production by Pachysolen tannophilus MTCC 1077- An economic & ecological venture. Indian J. of Biotechnology, 16, 444-456.

Bhatia, L. & Johri, S. (2018). Optimization of simultaneous saccharification and fermentation parameters for sustainable production of ethanol from wheat straw by Pichia stipitis NCIM 3498. Indian journal of Experimental Biology. 56.

Bhatia, L. (2019). Potential thermophilic enzymes for Bioethanol production. In: Sarangi, P.K. and Nanda, S. (Eds) "Biotechnology for Sustainable Energy and Products". IK International Publishing House Pvt. Ltd. New Delhi, pp. 129-148.

Bhatia, L., Bachetti, R., Ravindra, P. & Chandel, A. K. (2020). Third generation biorefineries: a sustainable platform for food, clean energy and nutraceuticals production. Biomass Conversion and Biorefinery. Springer. doi: 10.1007/s13399-020-00843-6

Bhatia, L., Garlapati, V. K. & Chandel, A. K. (2019). Scalable Technologies for Lignocellulosic Biomass Processing into Cellulosic Ethanol. In: Pogaku, R. (ed.), Horizons in Bioprocess Engineering. Springer International Publishing, (73-90). https://doi.org/10.1007/978-3-030-29069-6

Bhatia, L., Johri, S. & Ahmad, R. (2012). An economic and ecological perspective of ethanol production from renewable agro-waste- A review. Applied Microbiology & Biotechnology Express. 2 (65). doi:10.1186/2191-0855-2-65.

Bhatia, L., Singh, A., Chandel, A. K. & Singh, O. M. (2018). Biotechnological Advancements in cellulosic ethanol production. In: Singh, O. V., & Chandel, A. K. (eds.), Sustainable Biotechnology: Enzymatic Resources of Renewable

Energy. Chapter: 3, pp. 57-82. Springer International Publishing AG. doi: 10.1007/978-3-319-95480-6.

BIO (2016). Advancing the biobased economy: renewable chemical biorefinery commercialization, progress and market opportunities and beyond. https :// www.bio.org/advancing biobased economy renewable chemical biorefinery commercialization progress and market.

Bioamber Press release (2015). bioamber-now-shipping bio succinic acid to customers. https ://www.bio-amber .com/bioamber/en/news/2015.

Bozell, J. J. & Petersen, G. R. (2010). Technology development for the production of biobased products from biorefinery carbohydrates the US Department of Energy's "Top 10" revisited. Green Chem, 12, 539–554.

Bux, F. (2013). Potential biotechnological applications of microalgae: Biodiesel and value-added products. CRC Press, 1, 260.

Chandel, A. K., Bhatia, L., Garlapati, V. K., Roy, L., & Arora, A. (2017). Biofuel Policy in Indian Perspective: Socio-economic indicators and sustainable rural development. In: Chandel, A. K., & Sukumaran, R. K. (Eds.). Sustainable Biofuels Development in India. Springer International Publishing AG. Chapter: 19, pp.459-488. doi: 10.1007/978-3-319-50219-9_19.Switzerland,

Chandel, A. K., Garlapati, V. K., Singh, A.K., Antunes, F. A. F., & Silva, S. S. (2018). The path forward for lignocellulose biorefineries: bottlenecks, solutions, and perspective on commercialization. Bioresour Technol, 264, 370–381.

Chatterjee, C., Pong, F., & Sen, A. (2015). Chemical conversion pathways for carbohydrates. Green Chem, 17, 40–71.

Cherubini, F. (2010). The biorefinery concept: using biomass instead of oil for producing energy and chemicals. Energy Convers Manag, 51, 1412–1421.

Clark, J. H., & Deswarte, F. E. I. (2008). The biorefinery concept–an integrated approach. In: Clark, J. H., & Deswarte, F. E. I. (eds), Introduction to chemicals from biomass: (pp 1–20). Wiley, Chichester, https ://doi.org/10.1002/97804 70697 474

Clomburg, J. M. & Gonzalez, R. (2013). Anaerobic fermentation of glycerol: a platform for renewable fuels and chemicals. Trends Biotechnol, 31, 20–28.

de Jong, E., & Jungmeier, G. (2015). Biorefinery concepts in comparison to petrochemical refineries. In: Pandey, A., Hofer, R., Taherzadeh, M., Nampoothiri, M., & Larroche, C. (eds.), Industrial biorefineries and white biotechnology: (1st edn.) Amsterdam, Elsevier.

de Jong, E., Higson, A., Walsh, P., & Wellissch, M. (2013). Bio-based chemicals. value added products from biorefineries. In: IEA Bioenergy Task 42 report 2012. www.ieabioener gy.com/wp-content/uploads/2013/10/Task-42- Biobased Chemicals value-added products from-biorefineries.pdf.

Delhomme, C., Weuster-Botz, D. & Kuhn, F. E. (2009). Succinic acid from renewable resources as a C4 building-block chemical—a review of the catalytic possibilities in aqueous media. Green Chem, 11, 13–26.

Dietrich, K., Hernandez-Mejia, C., Verschuren, P., Rothenberg, G. & Shiju, N. R. (2017). One-pot selective conversion of hemicellulose to xylitol. Org Process Res Dev, 21, 165–170.

Direvo press release. (2013). http://www.direvo.com/uploads/media/DIREVO-PressRelease

Doan, Q. C., Moheimani, N. R., Mastrangelo, A. J. & Lewis, D. M. (2012). Microalgal biomass for bioethanol fermentation: Implications for hypersaline systems with an industrial focus. Biomass and bioenergy, 46, 79-88.

DuPont press release (2016). http://www.dupon t.com/products and services/ industrial biotechnology/press -releases/dupontadm announce platform technology for long sought-after molecule. html.

Eduardo, D. F. S. C. & Alberto, B. (2016). Bioethanol from microalgae and cyanobacteria: A review and technological outlook. Process Biochemistry, 51, 1833-1842.

Esposito, D. & Antonietti, M. (2015). Redefining biorefinery: the search for unconventional building blocks for materials. Chem Soc Rev, 44, 5821–5835.

Fernando, S., Adhikari, S., Chandrapal, C. & Murali, N. (2006). Biorefineries: current status, challenges and future direction. Energy Fuels, 20, 1727–1737.

Fonseca, G.G., Heinzle, E., Wittmann, C. & Gombert, A.K. (2008). The yeast Kluyveromyces marxianus and its biotechnological potential. Appl Microbiol Biotechnol, 79, 339–354.

G. F. (2015). Biochemicals press release. chemicals-announces-production. http://www.gfbio chemicals.com/ news/2015/07/gfbio -start -up/

Ghaffar, T., Irshad, M., Anwar, Z., Aqil, T., Zulifqar, Z., Tariq, A., Kamran, M., Ehsan, N. & Mehmood, S. (2014). Recent trends in lactic acid biotechnology: a brief review on production to purification. J Radiat Res Appl Sci, 7, 222–229.

Gil, S., Cuenca, N., Romero, A., Valverde, J. L. & Sáchez-Silva, L. (2014). Optimization of the synthesis procedure of microparticles containing gold for the selective oxidation of glycerol. Appl Catal A, 472, 11–20.

Hakkim, H. S., Raveendran, S., Pandey, A. & Parameswaran, B. (2020). Lignocellulosic bio-refinery approach for microbial 2,3-Butanediol production. Bioresource Technology, 302, doi: 10.1016/j.biortech.2020.122873

Hallenbeck, P. C. & Ghosh, D. (2009). Advances in fermentative biohydrogen production: the way forward. Trends Biotechnol, 27, 287–297. doi: 10.1016/j.tibtech.2009.02.004

Hendriks, A. T. W. M. & Zeeman, G. (2009). Pretreatments to enhance the digestibility of lignocellulosic biomass. Bioresour Technol, 100, 10–18.

Isikgor, F. H. & Becer, C. R. (2015). Lignocellulosic biomass: a sustainable platform for the production of bio-based chemicals and polymers. Poly Chem, 6, 4497–4559.

Jeffries, T.W., Grigoriev, I.V., Grimwood, J., Laplaza, J.M., Aerts, A., Salamov, A., Schmutz, J., Lindquist, E., Dehal, P., Shapiro, H., Jin, Y.S., Passoth, V. & Richardson, P.M. (2007). Genome sequence of the lignocellulose-bioconverting and xylose fermenting yeast Pichia stipitis. Nature Biotechnology, 25, 319-326.

Kamm, B. & Kamm, M. (2004). Principles of biorefineries. Appl Microbiol Biotechnol, 64, 137–145.

Kamm, B., Gruber, P. R. & Kamm, M. (2006). Biorefineries-industrial processes and products: Weinheim, Wiley-VCH Verlag GmbH & Co. K GaA.

Katiyar, R., Gurjar, B. R., Biswas, S., Pruthi, V., Nalin Kumar, N. & Kumar, P. (2017). Microalgae: An emerging source of energy-based bio-products and a solution for environmental issues. Renewable and Sustainable Energy Reviews, 72, 1083-1093.

Khan, M. I., Shin, J. H. & Kim, J. D. (2018). The promising future of microalgae: current status, challenges, and optimization of a sustainable and renewable industry for biofuels, feed, and other products. Microb Cell Fact, 17, 36.

Kildegaard, K. R., Wang, Z., Chen, Y., Nielsen, J. & Borodina, I. (2015). Production of 3-hydroxypropionic acid from glucose and xylose by metabolically engineered Saccharomyces cerevisiae. Metab Eng Commun, 2, 132–136.

Kumar, M. & Gayen, K. (2011). Developments in biobutanol production: new insights. Appl Energy, 88, 1999–2012. doi: 10.1016/j.apenergy.2010.12.055

Kumar, V., Ashok, S. & Park, S. (2013). Recent advances in biological production of 3-hydroxypropionic acid. Biotechnol Adv, 31, 945–961.

Lee, H. S., Vermaas, W. F. J. & Rittmann, B. E. (2010). Biological hydrogen production: prospects and challenges. Trends Biotechnol, 28, 262–271. doi: 10.1016/j.tibtech.2010.01.007

Lehtoma¨ki, A., Bj€ornsson, L. (2006). Two-stage anaerobic digestion of energy crops: methane production, nitrogen mineralisation and heavy metal mobilisation. Environ Technol, 27 (2), 209–218.

Li, H., Yang, S., Riisager, A., Pandey, A., Sangwan, R. S., Saravanamurugan, S. & Luque, R. (2016). Zeolite and zeotype-catalysed transformations of biofuranic compounds. Green Chem, 18, 5701–5735.

Liu, Z., Feist, A. M., Dragone, G. & Mussatto, S. I. (2019). Lipid and carotenoid production from wheat straw hydrolysates by different oleaginous yeasts. Journal of Cleaner Production, 249, 119308. doi: 10.1016/j.jclepro.2019.119308

Lugani, Y., Oberoi, S. & Sooch, B. S. (2017). Xylitol: a sugar substitute for patients of diabetes mellitus. World J Pharm Pharm Sci, 6,741–749.

Maeda, Y., Yoshino, T., Matsunaga, T., Matsumoto, M. & Tanaka, T. (2018). Marine microalgae for production of biofuels and chemicals. Current Opinion in Biotechnology, 50, 111–120.

Mariscal, R., Maireles-Torres, P., Ojeda, M., Sadaba, I. & Lopez Granados, M. (2016). Furfural: a renewable and versatile platform molecule for the synthesis of chemicals and fuels. Energy Environ Sci, 9, 1144–1189.

Mishra, B., Varjani, S. J. & Gottimukkala, K. S. V. (2019). Agro-Industrial By-Products in the Synthesis of Food Grade Microbial Pigments: An Eco-Friendly Alternative: Enzymes in Industrial Food Processing. In book: Green Bio-processes: doi: 10.1007/978-981-13-3263-0_13

Molino, A., Nanna, F., Ding, Y., Bikson, B. & Braccio, G. (2013). Biomethane production by anaerobic digestion of organic waste. Fuel, 103, 1003–1009. doi: 10.1016/j.fuel.2012.07.070

Naik, S. N., Goud, V. V., Rout, P. K. & Dalai, A. K. (2010). Production of first- and second-generation biofuels: a comprehensive review. Renew Sustain Energy Rev, 14, 578–597.

Nakagawa, Y., Tamura, M. & Tomishige, K. (2014). Catalytic materials for the hydrogenolysis of glycerol to 1,3-propanediol. J Mater Chem A, 2, 6688–6702.

Nanda, S., Azargohar, R., Dalai, A. K. & Kozinski, J. A. (2015). An assessment on the sustainability of lignocellulosic biomass for biorefining. Renew Sustain Energy Rev, 50, 925–941.

Nath, K. & Das, D. (2004). Improvement of fermentative hydrogen production: various approaches. Appl Microbiol Biotechnol, 65, 520–529.

Niesner, J., Jecha, D. & Stehl, K. P. (2013). Biogas upgrading technologies: state of art review in european region. Chem Eng Trans, 35, 517–522. doi:10.3303/CET1335086

NREL. https ://www.nrel.gov/working with us/re-biomass.html.

Octave, S. & Thomas, D. (2009). Biorefinery: toward an industrial metabolism. Biochimie, 91, 659–664.

Okoye, P. U. & Hameed. B. H. (2016). Review on recent progress in catalytic carboxylation and acetylation of glycerol as a byproduct of biodiesel production. Renew Sustain Energy Rev, 53, 558–574.

Ortiz, M. E., Bleckwedel, J., Raya, R. R. & Mozzi, F. (2013). Biotechnological and in situ food production of polyols by lactic acid bacteria. Appl Microbiol Biotechnol, 97, 4713–4726.

Pandey, A. (ed.) (2011). Biofuels: alternative feedstocks and conversion processes. Academic Press, Kidlington. ISBN-13: 978-0123850997

Pileidis, F. D. & Titirici, M. M. (2016). Levulinic acid biorefineries: new challenges for efficient utilization of biomass. Chem Sus Chem, 9, 562–582.

Ranalli, P. (2007). Improvement of crop plants for industrial end uses.

Redwood, M. D., Paterson-Beedle, M. & MacAskie, L. E. (2009). Integrating dark and light bio-hydrogen production strategies: towards the hydrogen economy. Rev Environ Sci Biotechnol, 8, 149–185. doi:10.1007/s11157-008-9144-9

Ribeiro, L. S., Delgado. J. J., de Melo Órfão, J. J. & Pereira, M. F. R. (2016). A one-pot method for the enhanced production of xylitol directly from hemicellulose (corncob xylan). RSC Adv, 6, 95320–95327.

Rodrigues, C. S., Villela, H. D. M., Martins, A. P., Marques, L. G., Colepicolo, P. & Tonon, A. P. (2015). Microalgae for economic applications: advantages and perspectives for bioethanol. Journal of Experimental Botany, 66, 4097-108.

Roquette/DSM (2008) Press release. https ://www.dsm.com/corpo rate/ about / business-entities/dsm biobased products and services/rever dia.html.

Rose, M. & Palkovits, R. (2011). Cellulose-based sustainable polymers: state of the art and future trends. Macromol Rapid Commun, 32, 1299–1311.

Rout, P. K., Nannaware, A. D, Prakash, O., Kalra, A. & Rajasekharan, R. (2016). Synthesis of hydroxymethylfurfural from cellulose using green processes: a promising biochemical and biofuel feedstock. Chem Eng Sci, 142, 318–346.

Ryckebosch, E., Drouillon, M. & Vervaeren, H. (2011). Techniques for transformation of biogas to biomethane. Biomass Bioenergy, 35, 1633–1645. doi: 10.1016/j.biombioe.2011.02.033

Sanchez, S., Bravo, V., Castro, E., Moya, A.J. & Camacho, F. (1999). Comparative study of the fermentation of D-glucose/D-xylose mixtures with Pachysolen tannophilus and Candida shehatae. Bioprocess Eng, 21, 525–532.

Santacesaria, E., Tesser, R., Di Serio, M., Casale, L. & Verde, D. (2010). New process for producing epichlorohydrin via glycerol chlorination. Ind Eng Chem Res, 49, 964–970.

Serrano-Ruiz, J. C. & Dumesic, J. A. (2011). Catalytic routes for the conversion of biomass into liquid hydrocarbon transportation fuels. Energy Environ Sci, 4, 83–99.

Sims, R., Taylor, M., Jack, S. & Mabee, W. (2008). From 1st to 2nd generation bio fuel technologies: an overview of current industry and RD&D activities. IEA Bioenergy, 1–124.

Sudhakar, K. & Premalatha, M. (2015). Characterization of Micro Algal Biomass Through FTIR/TGA /CHN Analysis: Application to Scenedesmus sp. Energy Sources. Recovery, Utilization, and Environmental Effects, 37, 2330-2337.

Sudhakar, T., Li, T. & Gonzalez, M. A. (2018). An overview of biorefinery derived platform chemicals from a cellulose and hemicellulose biorefinery. Clean Technologies and Environmental Policy, 20, 1615–1630. https://doi.org/10.1007/s10098-018-1568-5

Tabah, B., Varvak, A., Pulidindi, I.N., Foran, E., Baninb, E. & Gedanken, A., (2016). Production of 1,3-propanediol from glycerol via fermentation by Saccharomyces cerevisiae. Green Chem, 18, 4657. doi: 10.1039/c6gc00125d.

Takkellapati, S., Li, T. & Gonzalez, M.A. (2018). An overview of biorefinery-derived platform chemicals from a cellulose and hemicellulose biorefinery. Clean Technologies and Environmental Policy, 20, 1615–1630. https://doi.org/10.1007/s10098-018-1568-5

Valin, H., Sands, R. D., van der Mensbrugghe, D., Nelson, G. C., Ahammad, H., Blanc, E. & Willenbocke. I. D. (2013). The future of food demand: understanding differences in global economic models. Agric Econ, 45, 51–67.

Verma, S., Baig, R. B. N., Nadagouda, M. N., Len, C. & Varma, R. S. (2017). Sustainable pathway to furanics from biomass via heterogeneous organo-catalysis. Green Chem, 19, 164–168.

Werpy, T., Petersen, G., Aden, A., Bozell, J., Holladay, J., White, J. & Manheim, A. (2004). In: Werpy T, Petersen G (eds) Top value-added chemicals from biomass—vol. 1: results of screening for potential candidates from sugars and synthesis gas. Pacific Northwest National Laboratory, National Renewable Energy Laboratory and Department of Energy, Washington, DC

Whited, G. M., Feher, F. J., Benko, D. A., Cervin, M. A., Chotani, G. K., McAuliffe, J. C., LaDuca, R. J., Ben-Shoshan, E. A. & Sanford, K. J. (2010). Technology update: development of a gas-phase bioprocess for isoprenemonomer production using metabolic pathway engineering. Ind Biotechnol, 6, 152–163.

Wu, Y-D., Xue, C., Chen, L-J. & Bai, F-W. (2013). Effect of zinc supplementation on acetone– butanol–ethanol fermentation by Clostridium acetobutylicum. J Biotechnol, 165, 18–21. doi:10. 1016/j.jbiotec.2013.02.009

Zhang, J., Li, J., Wu, S. & Liu, Y. (2013). Advances in the catalytic production and utilization of sorbitol. Ind Eng Chem Res, 52, 11799–11815.

Zhang, X., Durndell, L. J., Isaacs, M. A., Parlett, C. M. A., Lee, A. F. & Wilson, K. (2016). Platinum-catalyzed aqueous-phase hydrogenation of d-glucose to d-sorbitol. ACS Catal, 6, 7409–7417.

Zheng, P., Wereath, K., Sun, J. B., van den Heuvel, J. & Zeng, A. (2006). Overexpression of genes of the dha regulon and its effects on cell growth, glycerol fermentation to 1,3-propanediol and plasmid stability in Klebsiella pneumoniae. Process Biochem, 41, 2160–2169.

Chapter-9

INNOVATIONS IN FERMENTER ENGINEERING: EXPLORING NOVEL FERMENTER DESIGNS AND SPECIALIZED TYPES OF ENHANCED FERMENTATION PROCESSES

*Tanushree Sarkar and Harit Jha**

Department of Biotechnology, Guru Ghasidas Vishwavidyalaya (A Central University), Bilaspur, Chhattisgarh, 495009, INDIA

**Corresponding author: Email ID: harit74@yahoo.co.in*

Abstract

In the field of fermentation, innovative solutions in fermenter design have emerged to address challenges and optimize process efficiency. This review highlights the significance of innovative fermenter design and its special types in enhancing product yield, quality, and scalability. Each type of fermenter has its advantages and limitations, making it suitable for specific applications. Understanding the characteristics and operational considerations of different fermenter types is crucial for selecting the appropriate system and optimizing fermentation processes in various industries. The review also provides an overview of the key advancements and their impact on fermentation processes. Innovative solutions in fermenter design have revolutionized the way fermentation processes are carried out. Traditional fermenter designs have been improved upon, and new designs have been introduced to overcome limitations and optimize performance. These innovations have paved the way for improved product yields, reduced production costs, and enhanced process control.

Keywords: Fermenter, *Fermenter design, Fermenter operation, Types of fermenter*

9.1. Introduction

Fermentation is a vital process that has been employed by humans for centuries in various industries, ranging from food and beverage production to pharmaceuticals and biofuel manufacturing. The use of fermenters, also known as bioreactors, has revolutionized the efficiency and scale of fermentation processes, enabling the production of a wide array of valuable products. In this comprehensive review, we delve into the world of fermenters, exploring their types, applications, and advancements that have shaped the field of fermentation technology (Patel et al., 2023).

Fermenters, as specialized vessels designed for carrying out fermentation processes, have evolved significantly over time. They provide a controlled environment for microbial, enzymatic, or cellular activities, allowing the conversion of raw materials into desired products (Sharma et al., 2020). The design and operation of fermenters play a crucial role in achieving optimal yields, productivity, and product quality, making them indispensable tools in the biotechnology industry (Anand et al., 2013).

The primary objective of this review is to provide an in-depth analysis of different types of fermenters, including their design principles, operating parameters, and applications. We will explore the historical development of fermenters, starting from the early rudimentary vessels to the sophisticated, high-tech bioreactors utilized today. By examining their evolution, we can appreciate the milestones that have propelled the field forward and understand the rationale behind the current fermenter designs.

Fermenters are categorised based on various criteria, including their mode of operation, scale, and the type of organisms or processes they support (Jha and Patil., 2011). Batch fermenters, where a finite amount of substrate is added and the process is carried out until completion, will be discussed in detail. Continuous fermenters, which enable a continuous supply of nutrients and product removal, will also be explored. Additionally, we will explore types of fermenters specific to particular applications, such as solid-state fermenters, submerged fermenters, and airlift fermenters (Yang and Sha., 2019).

The chapter will highlight the critical factors that influence fermenter performance, such as temperature, pH, agitation, aeration, and substrate concentration. We will discuss the engineering considerations and design parameters involved in optimizing these factors to enhance fermentation efficiency and productivity. Furthermore, we will examine the instrumentation and control systems employed in modern fermenters, which enable real-time monitoring and precise regulation of crucial process variables.

The applications of fermenters are vast and encompass diverse industries. We will explore their role in the production of various commodities, including alcoholic beverages, dairy products, antibiotics, enzymes, and biofuels. Furthermore, we will delve into emerging areas where fermenters are being employed, such as the production of biopharmaceuticals, nutraceuticals, and bio-based materials (Mears et al., 2017). By examining these applications, we can appreciate the far-reaching impact of fermenters in shaping our daily lives and fostering sustainable development.

Lastly, recent advancements in fermenter technology that have revolutionized the field will be discussed. This includes the integration of automation, artificial intelligence, and data analytics in fermenter systems, enabling enhanced process control, yield optimization, and cost reduction (Yang et al., 2023). We will also explore the adoption of novel bioreactor designs, such as single-use and microfluidic systems, which offer advantages in terms of scalability, flexibility, and ease of operation.

9.2. Fermenter Design and Operations

The design and operation of fermenters play a critical role in optimizing product yield, quality, and process efficiency. The success of a fermentation process heavily relies on the design and operation of fermenters (Bhargav et al., 2008). This subsection highlights the critical role of fermenter design in ensuring optimal process outcomes and explores the significance of efficient operational strategies. Fundamentals of fermenter design include:

9.2.1 Scaling-Up: From Laboratory to Industrial Scale

Scaling-up fermentation processes is a complex task. This section discusses the challenges and considerations associated with transitioning from laboratory-scale to large-scale fermenters, emphasizing the importance of maintaining process consistency (Schmidt., 2005).

9.2.2 Design Considerations and Selection Criteria

An effective fermenter design should address key factors such as vessel geometry, impeller type, vessel material, and sterilization methods. This subsection explores the design considerations and selection criteria for fermenters, focusing on aspects that influence mixing, heat transfer, and aeration (Gaikwad et al., 2023).

9.2.3 Material Selection and Sterility

Fermenters must be constructed from materials that are compatible with the process requirements and ensure sterility. This section discusses the commonly used materials for fermenter construction, their advantages, limitations, and the importance of maintaining sterile conditions (Geigert., 2023).

9.2.4 Agitation and Aeration Systems

Efficient agitation and aeration are crucial for maintaining optimal microbial growth and product formation. This subsection explores various agitation mechanisms, aeration strategies, and their impact on process performance (Mohseni et al., 2023).

9.2.5 Heat Transfer and Cooling Systems

Heat transfer is essential for controlling and dissipating heat generated during fermentation. This section highlights heat transfer mechanisms and cooling systems utilized in fermenter design, emphasizing the importance of maintaining optimal temperature conditions (Tusset et al., 2022).

9.2.6 pH and Temperature Control

Maintaining optimal pH and temperature conditions is critical for microbial growth and product synthesis (Manaf et al., 2022). This subsection discusses the challenges associated with pH and temperature control in fermenters and explores different control strategies employed in practice.

9.3. Special Purpose Fermenters

9.3.1. Photobioreactor

Photobioreactors (PBRs) (Figure 9.1) hold significant importance in the field of biotechnology and sustainable development. These specialized systems provide a controlled environment for the cultivation of photosynthetic microorganisms, such as algae and cyanobacteria, harnessing their ability to convert sunlight and carbon dioxide into valuable products (Sathinathan et al., 2023). The importance of photobioreactors lies in their ability to offer a scalable and sustainable solution for various applications. They enable efficient production of biofuels, including biodiesel, bioethanol, and biogas, offering renewable alternatives to fossil fuels (Nwoba et al., 2019).

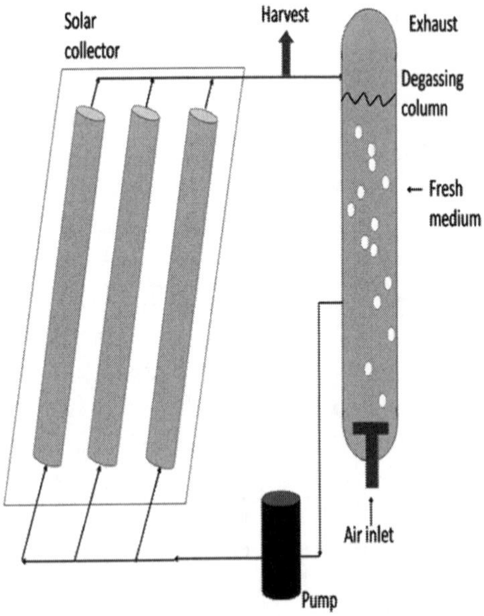

Figure 9.1. Photobioreactor

Photobioreactors also serve as platforms for the synthesis of high-value products such as omega-3 fatty acids, pigments, and pharmaceuticals. Moreover, they play a crucial role in wastewater treatment and bioremediation, as they can remove pollutants and capture carbon dioxide. The versatility and potential of photobioreactors in addressing global challenges, such as energy security, environmental sustainability, and resource scarcity, make them an essential tool for the development of a bio-based economy and a greener future (Senatore et al., 2021).

9.3.2. Solid-state Fermentation

Solid-state fermentation (SSF) and the design of solid-state fermenters hold great importance in various industries, particularly in the production of enzymes, organic acids, bioactive compounds, and microbial biomass from solid substrates. Unlike submerged fermentation, SSF utilizes solid materials as the substrate, offering several advantages such as lower water requirement, reduced energy consumption, and ease of product recovery (Chilakamarry et al., 2022)

The importance of solid-state fermenters lies in their ability to provide a suitable environment for microbial growth and metabolic activity within the solid matrix, facilitating efficient nutrient utilization and product formation. The design of solid-state fermenters is crucial to create optimal conditions for microbial colonization, oxygen and heat transfer, and moisture control (Kumar et al., 2021). Various factors such as the choice of solid substrate, particle size, moisture content, aeration, and temperature play a vital role in designing solid-state fermenters.

Additionally, the selection of appropriate microbial strains and the optimization of process parameters are essential considerations for achieving desired product yields and quality in solid-state fermentation. The importance and proper design of solid-state fermenters contribute to the development of sustainable bioprocesses, resource utilization, and the production of valuable bioproducts from solid waste materials (Singhania et al., 2009).

9.3.3. Bubble Column Fermenter

Bubble column fermenters are of great importance in biotechnology and industrial fermentation processes. These fermenters utilize the principle of gas-liquid interaction, where gas bubbles provide mixing, aeration, and mass transfer within the liquid phase (Chen et al., 2015). The importance of bubble column fermenters lies in their ability to facilitate the cultivation of microorganisms under controlled conditions and to optimize the production of various metabolites, enzymes, and biopolymers. The design of bubble column fermenters is critical for achieving efficient gas-liquid contacting and maintaining optimal conditions for microbial growth (González-Gloria et al., 2022).

Key design parameters include the column height, diameter, liquid volume, sparger design, and bubble size control. The column height and diameter influence the liquid holdup, residence time, and gas holdup, which, in turn, affect the mixing efficiency and oxygen transfer rate (Laguillaumie et al., 2022). The sparger design is crucial for generating uniform bubble distribution and promoting gas-liquid contact. Additionally, the control of bubble size helps optimize mass transfer and prevent excessive shear stress on the microbial cells. The design of bubble column fermenters aims to create an optimal environment for microbial growth, ensuring efficient nutrient utilization, and enhancing the productivity and yield of target products (Sitaraman et al., 2022).

9.3.4. Airlift Bioreactor (ALB)

Airlift bioreactors (Figure 9.2) play a significant role in various biotechnological applications, making them of utmost importance in the field. These bioreactors operate based on the principle of gas-liquid circulation, utilizing the buoyancy of gas bubbles to create upward flow and promote mixing and aeration within the system (Negi et al., 2023).

The importance of airlift bioreactors lies in their ability to provide a gentle and efficient environment for the cultivation of sensitive organisms, including animal and plant cells, as well as microorganisms (Gholami et al., 2020). Airlift bioreactors offer advantages such as low shear stress, enhanced mass transfer, and reduced damage to fragile cells, making them ideal for applications such as cell culture, bioactive compound synthesis, and wastewater treatment. The design of airlift bioreactors is crucial to optimize their performance. Key design parameters include the riser and downcomer dimensions, gas and liquid flow rates, and the positioning of gas spargers and liquid distributors (Mutaf and Oncel., 2023).

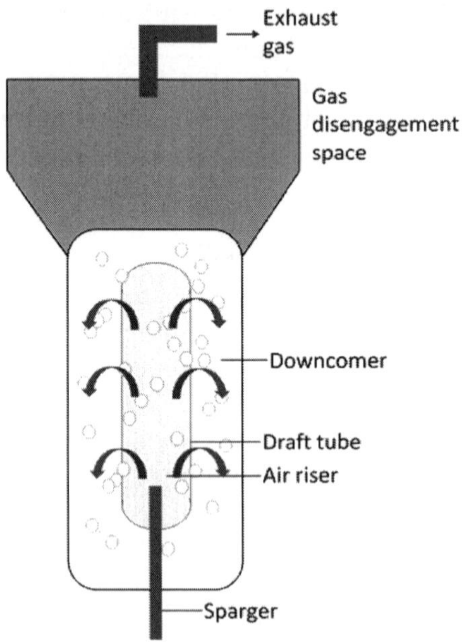

Figure 9.2. Airlift Bioreactor

The design aims to achieve efficient gas-liquid contacting, adequate mixing, and appropriate oxygen transfer rates to support optimal cell growth and product formation. Airlift bioreactors provide a versatile and reliable platform for a wide range of bioprocesses, contributing to advancements in biotechnology, pharmaceuticals, and environmental engineering (Merchuk et al., 2003).

9.3.5. Packed Bed Bioreactors

Packed bed bioreactors (Figure 9.3) are of paramount importance in various bioprocesses, offering advantages in terms of high cell density, stability, and ease of operation. These bioreactors consist of a fixed bed of solid support material, such as porous beads or fibrous matrices, where microbial cells are immobilized (Meuwly et al., 2007). The importance of packed bed bioreactors lies in their ability to provide an efficient environment for cell growth and metabolism while facilitating the continuous flow of substrate through the packed bed (Warnock et al., 2005).

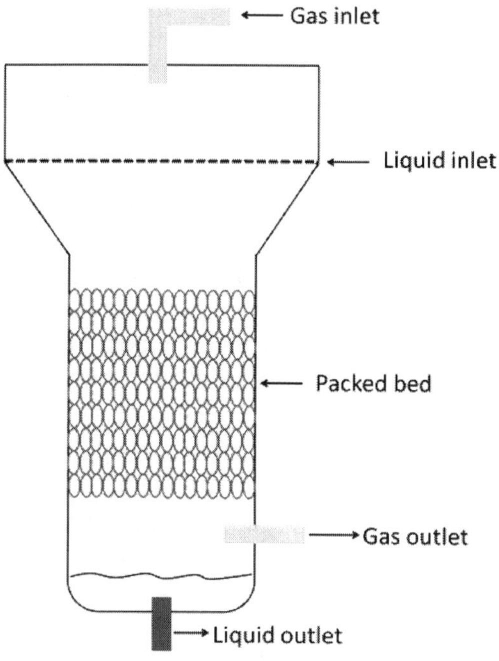

Figure 9.3. Packed Bed Bioreactor

The design of packed bed bioreactors is critical to ensure optimal mass transfer, substrate utilization, and product formation. Key design parameters include the selection of suitable support material, determination of bed height and diameter, and optimization of flow rates. The choice of support material affects factors such as cell attachment, mechanical stability, and nutrient diffusion, which ultimately impact overall reactor performance (Mitchell **et al., 2023).** Proper design considerations help maintain a uniform flow of substrate, promote efficient nutrient uptake, and facilitate effective removal of metabolites.

Packed bed bioreactors find applications in various industries, including enzyme production, wastewater treatment, and continuous fermentation processes, due to their advantages in terms of high productivity, long operational lifespan, and potential for scale-up (Shareefdeen et al., 2022). The importance and appropriate design of packed bed bioreactors contribute to efficient and stable bioprocesses, allowing for the continuous production of valuable products in a controlled and sustainable manner.

9.3.6. Fluidized Bed Reactor

Fluidized bed reactors are of significant importance in various biotechnological and chemical processes. These reactors employ the principle of fluidization, where solid particles are suspended and mobilized by an upward flow of fluid, typically air or gas (Luo et al., 2022). The importance of fluidized bed reactors lies in their ability to provide excellent mixing, heat transfer, and mass transfer characteristics. Fluidized bed reactors offer advantages such as high surface area-to-volume ratio, efficient nutrient utilization, and enhanced mass transfer rates, making them suitable for a wide range of applications (Manzoor et al., 2021).

They are commonly used in the production of enzymes, bioactive compounds, and microbial biomass. The design of fluidized bed reactors is crucial to ensure optimal performance. Key design parameters include particle size, bed height, fluid velocity, and the choice of solid particles. The selection of appropriate particle size allows for effective fluidization and particle movement, while the bed height and fluid velocity determine the residence time and mixing within the reactor (Sharma et al., 2021). The choice of solid particles affects factors such as heat transfer, nutrient diffusion, and biomass growth.

Proper design considerations help achieve efficient fluidization, minimize particle attrition, and optimize reactor performance. Fluidized bed reactors play a vital role in enabling sustainable and efficient bioprocesses, contributing to the production of valuable products and the advancement of various industries (McKenna et al., 2019).

9.3.7. Hollow Fiber Cartridges

Hollow fiber cartridge fermenters combine the advantages of hollow fiber cartridges with the functionality of fermenters, making them essential in the field of bioprocessing. These fermenters utilize a bundle of hollow fibers with a porous membrane as the growth substrate and separation medium for microbial cultures. The importance of hollow fiber cartridge fermenters lies in their ability to provide a high surface area for cell growth and efficient mass transfer while allowing for continuous separation and harvest of products (Wan et al., 2021).

The design of hollow fiber cartridge fermenters is crucial to optimize their performance. Key design parameters include the selection of appropriate membrane material, pore size, and fiber diameter, as well as the configuration of the cartridge and flow dynamics within the system. The membrane material and pore size determine the permeability and selectivity of the cartridge, enabling the separation of cells and metabolites while allowing for nutrient and waste exchange. The fiber diameter affects the available surface area for cell attachment and growth. Proper design considerations ensure optimal nutrient utilization, waste removal, and product yield (Zrellia et al., 2023).

Hollow fiber cartridge fermenters find applications in various bioprocesses, such as continuous fermentation, cell culture, and bioproduction of valuable compounds. Their importance and appropriate design contribute to improved productivity, process efficiency, and product quality in biotechnology and pharmaceutical industries (Eghbali et al., 2016).

9.3.8. Rocking Bag Bioreactors

Rocking bag bioreactors are of significant importance in the field of bioprocessing, offering unique advantages for the cultivation of various cell types. These bioreactors utilize a flexible bag that is rocked back and forth, creating gentle agitation and promoting efficient mixing and gas exchange within the culture (Eibl and Eibl., 2009). The importance of rocking bag bioreactors lies in their ability to provide a controlled and optimized environment for the growth and production of cells, including mammalian cells, stem cells, and microbial cultures (Sharma et al., 2022).

The rocking motion helps maintain homogeneous conditions, minimizes shear stress, and enhances oxygen and nutrient availability throughout the culture. The design of rocking bag bioreactors is critical to ensure proper mixing, sterility, and scalability. Key design parameters include the size and shape of the bag, the rocking frequency and angle, and the control systems for temperature, pH, and dissolved oxygen. The bag material should be biocompatible, sterile, and suitable for gas exchange, while the rocking mechanism should provide gentle and consistent agitation (Murthy et al., 2023).

Proper design considerations help ensure optimal growth, viability, and productivity of the cells within the bioreactor. Rocking bag bioreactors find applications in various fields, including biopharmaceutical production, regenerative medicine, and vaccine development. Their importance and appropriate design contribute to improved cell culture performance, reduced operating costs, and increased process efficiency (Kheirkhah et al., 2023).

9.3.9. Stem Cell Bioreactor

Stem cell bioreactors play a crucial role in stem cell research and regenerative medicine, making them of paramount importance in the field. These bioreactors provide a controlled environment for the expansion, differentiation, and manipulation of stem cells, enabling their potential therapeutic applications (Li et al., 2023). The importance of stem cell bioreactors lies in their ability to mimic the physiological conditions and microenvironment necessary for stem cell growth and differentiation (Kweon et al., 2023). They provide precise control over factors such as temperature, oxygen tension, nutrient supply, and mechanical stimulation, which are critical for regulating stem cell behaviour and fate.

The design of stem cell bioreactors is critical to create optimal conditions for stem cell culture (Yeatts et al., 2013). Key design parameters include the choice of bioreactor type, such as spinner flasks, stirred tank bioreactors, or perfusion systems, as well as the selection of scaffold materials and surface coatings for cell attachment and differentiation. The bioreactor design should also allow for scalability, ease of cell monitoring, and efficient nutrient and waste exchange (King and Miller., 2007).

Proper design considerations facilitate the expansion and differentiation of stem cells in a controlled and reproducible manner, leading to improved cell viability, functionality, and yield. Stem cell bioreactors have revolutionized stem cell research, offering the potential for large-scale production of cells for therapeutic purposes, tissue engineering, and drug discovery. Their importance and appropriate design contribute to advancing the field of regenerative medicine and realizing the potential of stem cell-based therapies (Petry and Salzig., 2021).

9.4. Conclusion

In conclusion, fermenters have undergone significant innovations and advancements in their design to meet the growing demands of biotechnological applications. The integration of innovative technologies and engineering principles has revolutionized fermenter design, enabling improved control over key parameters such as temperature, pH, oxygen transfer, and mixing. These advancements have resulted in enhanced productivity, yield, and scalability in various fermentation processes.

Furthermore, the development of specialized fermenter types, such as photobioreactors, solid-state fermenters, bubble column fermenters, packed bed bioreactors, and stem cell bioreactor, has opened up new avenues for diverse applications in industries ranging from energy and pharmaceuticals to wastewater treatment and bioremediation. Each fermenter type offers distinct advantages and addresses specific process requirements, demonstrating the importance of tailoring the design to suit the desired outcomes. Overall, the continuous innovation and development in fermenter design and the availability of various types of fermenters have significantly contributed to the advancement of biotechnology, promoting sustainable practices, and unlocking the potential for a wide range of bioprocesses and bioproducts.

References

Anand, B., Sudha, S., Keshaw, A., & Harit, J. (2013). Value added products from agrowaste. Recent Research in Science and Technology, 5(2).

Bhargav, S., Panda, B. P., Ali, M., & Javed, S. (2008). Solid-state fermentation: an overview. Chemical and Biochemical Engineering Quarterly, 22(1), 49-70.

Chen, J., Gomez, J. A., Höffner, K., Barton, P. I., & Henson, M. A. (2015). Metabolic modeling of synthesis gas fermentation in bubble column reactors. Biotechnology for biofuels, 8(1), 1-12.

Chilakamarry, C. R., Sakinah, A. M., Zularisam, A. W., Sirohi, R., Khilji, I. A., Ahmad, N., & Pandey, A. (2022). Advances in solid-state fermentation for bioconversion of agricultural wastes to value-added products: Opportunities and challenges. Bioresource technology, 343, 126065.

Eghbali, H., Nava, M. M., Mohebbi-Kalhori, D., & Raimondi, M. T. (2016). Hollow fiber bioreactor technology for tissue engineering applications. The International journal of artificial organs, 39(1), 1-15.

Eibl, R., & Eibl, D. (2009). Application of disposable bag bioreactors in tissue engineering and for the production of therapeutic agents. Bioreactor systems for tissue engineering, 183-207.

Gaikwad, V., Panghal, A., Jadhav, S., Kundu, S., Singh, N., & Chhikara, N. (2023). Design Schematics, Operational Characteristics and Process Applications of Bioreactors. Microbes in the Food Industry, 217-247.

Geigert, J. (2023). Upstream Production of the Biopharmaceutical Drug Substance. In The Challenge of CMC Regulatory Compliance for Biopharmaceuticals (pp. 231-270). Cham: Springer Nature Switzerland.

Gholami, F., Zinatizadeh, A. A., Zinadini, S., McKay, T., & Sibali, L. (2020). An innovative jet loop-airlift bioreactor for simultaneous removal of carbon and nitrogen from soft drink industrial wastewater: Process performance and kinetic evaluation. Environmental Technology & Innovation, 19, 100772.

González-Gloria, K. D., Rodríguez-Jasso, R. M., Saxena, R., Sindhu, R., Ali, S. S., Singhania, R. R., & Ruiz, H. A. (2022). Bubble column bioreactor design and evaluation for bioethanol production using simultaneous saccharification and fermentation strategy from hydrothermally pretreated lignocellulosic biomass. Biochemical Engineering Journal, 187, 108645.

Jha, H., & Patil, M. (2011). Biopulping of sugarcane bagasse using manganese peroxidase from Penicillium oxalicum isolate-1. Romanian Biotechnological Letters, 16(6), 6809-6819.

Kheirkhah, T., Neubauer, P., & Junne, S. (2023). Controlling Aspergillus niger morphology in a low shear-force environment in a rocking-motion bioreactor. Biochemical Engineering Journal, 195, 108905.

King, J. A., & Miller, W. M. (2007). Bioreactor development for stem cell expansion and controlled differentiation. Current opinion in chemical biology, 11(4), 394-398.

Kumar, V., Ahluwalia, V., Saran, S., Kumar, J., Patel, A. K., & Singhania, R. R. (2021). Recent developments on solid-state fermentation for production of microbial secondary metabolites: Challenges and solutions. Bioresource Technology, 323, 124566.

Kweon, S., Kim, S., & Baek, E. J. (2023). Current status of red blood cell manufacturing in 3D culture and bioreactors. Blood research, 58(S1), S46-S51.

Laguillaumie, L., Rafrafi, Y., Moya-Leclair, E., Delagnes, D., Dubos, S., Spérandio, M., & Dumas, C. (2022). Stability of ex situ biological methanation of H2/CO2 with a mixed microbial culture in a pilot scale bubble column reactor. Bioresource Technology, 354, 127180.

Li, J., Wu, Y., Yao, X., Tian, Y., Sun, X., Liu, Z., & Wu, C. (2023). Preclinical Research of Stem Cells: Challenges and Progress. Stem Cell Reviews and Reports, 1-15.

Luo, H., Wang, X., Liu, X., Wu, X., Shi, X., & Xiong, Q. (2022). A review on CFD simulation of biomass pyrolysis in fluidized bed reactors with emphasis on particle-scale models. Journal of Analytical and Applied Pyrolysis, 162, 105433.

Manaf, S. F. A., Luthfi, A. A. I., Jahim, J. M., Harun, S., Tan, J. P., & Shah, S. S. M. (2022). Sequential detoxification of oil palm fronds hydrolysate with coconut shell activated charcoal and pH controlled in bioreactor for xylitol production. Chemical Engineering Research and Design, 179, 90-106.

Manzoor, S., Tatum, J., Wani, O. B., & Bobicki, E. R. (2023). Comminution of carbon particles in a fluidized bed reactor: A review. Minerals Engineering, 195, 108026.

McKenna, T. F. (2019). Condensed mode cooling of ethylene polymerization in fluidized bed reactors. Macromolecular Reaction Engineering, 13(2), 1800026.

Mears, L., Feldman, H., Falco, F. C., Bach, C., Wu, M., Nørregaard, A., & Gernaey, K. V. (2017). Continuous fermentation for biopharmaceuticals?. Continuous Manufacturing of Pharmaceuticals, 227-246.

Merchuk, J. C. (2003). Airlift bioreactors: review of recent advances. The Canadian Journal of Chemical Engineering, 81(3-4), 324-337.

Meuwly, F., Ruffieux, P. A., Kadouri, A., & Von Stockar, U. (2007). Packed-bed bioreactors for mammalian cell culture: bioprocess and biomedical applications. Biotechnology Advances, 25(1), 45-56.

Mitchell, D. A., Ruiz, H. A., & Krieger, N. (2023). A Critical Evaluation of Recent Studies on Packed-Bed Bioreactors for Solid-State Fermentation. Processes, 11(3), 872.

Mohseni, S., Khoshfetrat, A. B., Rahbarghazi, R., Khodabakhshaghdam, S., & Kaleybar, L. S. (2023). Influence of shear force on ex vivo expansion of hematopoietic model cells in a stirred tank bioreactor. Journal of Biological Engineering, 17(1), 1-10.

Murthy, H. N., Joseph, K. S., Paek, K. Y., & Park, S. Y. (2023). Bioreactor systems for micropropagation of plants: present scenario and future prospects. Frontiers in Plant Science, 14, 1159588.

Mutaf, T., & Oncel, S. S. (2023). Bubble column and airlift bioreactor systems for animal cell culture applications. Asia-Pacific Journal of Chemical Engineering, 18(1), e2872.

Negi, B. B., Aliveli, M., Behera, S. K., Das, R., Sinharoy, A., Rene, E. R., & Pakshirajan, K. (2023). Predictive modelling and optimization of an airlift bioreactor for selenite removal from wastewater using artificial neural networks and particle swarm optimization. Environmental Research, 219, 115073.

Nwoba, E. G., Parlevliet, D. A., Laird, D. W., Alameh, K., & Moheimani, N. R. (2019). Light management technologies for increasing algal photobioreactor efficiency. Algal research, 39, 101433.

Patel, A. K., Dong, C. D., Chen, C. W., Pandey, A., & Singhania, R. R. (2023). Production, purification, and application of microbial enzymes. In Biotechnology of microbial enzymes (pp. 25-57). Academic Press.

Petry, F., & Salzig, D. (2021). Impact of bioreactor geometry on mesenchymal stem cell production in stirred-tank bioreactors. Chemie Ingenieur Technik, 93(10), 1537-1554.

Sathinathan, P., Parab, H. M., Yusoff, R., Ibrahim, S., Vello, V., & Ngoh, G. C. (2023). Photobioreactor design and parameters essential for algal cultivation using industrial wastewater: A review. Renewable and Sustainable Energy Reviews, 173, 113096.

Schmidt, F. R. (2005). Optimization and scale up of industrial fermentation processes. Applied microbiology and biotechnology, 68, 425-435.

Senatore, V., Buonerba, A., Zarra, T., Oliva, G., Belgiorno, V., Boguniewicz-Zablocka, J., & Naddeo, V. (2021). Innovative membrane photobioreactor for sustainable CO2 capture and utilization. Chemosphere, 273, 129682.

Shareefdeen, Z., & Qasim, M. (2022). A Critical Remark on the Applications of Gas-Phase Biofilter (Packed-Bed Bioreactor) Models in Aqueous Systems. Bioengineering, 9(11), 657.

Sharma, M., Mohapatra, T., & Ghosh, P. (2021). Hydrodynamics, mass and heat transfer study for emerging heterogeneous Fenton process in multiphase fluidized-bed reactor system for wastewater treatment—A review. Chemical Engineering Research and Design, 171, 48-62.

Sharma, R., Garg, P., Kumar, P., Bhatia, S. K., & Kulshrestha, S. (2020). Microbial fermentation and its role in quality improvement of fermented foods. Fermentation, 6(4), 106.

Sharma, R., Harrison, S. T., & Tai, S. L. (2022). Advances in bioreactor systems for the production of biologicals in mammalian cells. ChemBioEng Reviews, 9(1), 42-62.

Singhania, R. R., Patel, A. K., Soccol, C. R., & Pandey, A. (2009). Recent advances in solid-state fermentation. Biochemical Engineering Journal, 44(1), 13-18.

Sitaraman, H., Hassanaly, M., Parra-Alvarez, M., Rahimi, M., & Stickel, J. (2022). Impact of Variable Gas Mixtures on Bubble Size Distribution and Mass Transfer in Gas Fermentation Reactors (No. NREL/PR-2C00-84546). National Renewable Energy Lab.(NREL), Golden, CO (United States).

Tusset, A. M., Inacio, D., Fuziki, M. E., Costa, P. M., & Lenzi, G. G. (2022). Dynamic analysis and control for a bioreactor in fractional order. Symmetry, 14(8), 1609.

Wan, C. F., Yang, T., Lipscomb, G. G., Stookey, D. J., & Chung, T. S. (2021). Design and fabrication of hollow fiber membrane modules. In Hollow Fiber Membranes (pp. 225-252). Elsevier.

Warnock, J. N., Bratch, K., & Al-Rubeai, M. (2005). Packed bed bioreactors. In Bioreactors for Tissue Engineering: Principles, Design and Operation (pp. 87-113). Dordrecht: Springer Netherlands.

Yang, C. T., Kristiani, E., Leong, Y. K., & Chang, J. S. (2023). Big Data and Machine Learning Driven Bioprocessing-Recent trends and critical analysis. Bioresource Technology, 128625.

Yang, Y., & Sha, M. (2019). A beginner's guide to bioprocess modes–batch, fed-batch, and continuous fermentation. Enfield, CT: Eppendorf Inc, 408, 1-16.

Yeatts, A. B., Choquette, D. T., & Fisher, J. P. (2013). Bioreactors to influence stem cell fate: augmentation of mesenchymal stem cell signaling pathways via dynamic culture systems. Biochimica et Biophysica Acta (BBA)-General Subjects, 1830(2), 2470-2480.

Zrellia, A., Debayab, J., Doucourec, A., & Chaouachid, B. (2023). Applications of Helical Versus Straight Hollow Fiber Membranes: A. Engineering and Technology Journal, 41(03), 492-502.

Chapter-10

ALGAE EDUCED INDUSTRIAL ENZYMES PRODUCTION: A SUSTAINABLE WAY FORWARD?

Dipankar Ghosh[1], Shreyasmita Guha[1], Dona Roy[1], Ayan Samanta[1], Kamalendu De[1], Annesha Chakraborty[1], Supriyo Mal[1], Ankita Gole[1] and Rishav Kar[2]*

[1]Microbial Engineering and Algal Biotechnology Laboratory, Department of Biosciences, JIS University, Kolkata, West Bengal, India.

[2]Department of Microbiology, Ramakrishna Mission Vivekananda Centenary College, Rahara, Khardaha, West Bengal, India.

**Corresponding author Email: d.ghosh@jisuniversity.ac.in or dghosh.jisuniversity@gmail.com*

Abstract

Industrial enzyme production has been predominated by the involvement of bacterial and fungal enzymes in global perspective since last few decades. Bacteria and fungi-based enzymes are employed for scaling up due its higher specificity and catalytic efficiencies with minimal by-product generation. Industrially feasible bacterial and fungal enzymes include cellulase, amylase, protease, lipase, glucose oxidase, pectinase, and tannase. However, these bacterial and fungal industrial enzymes are not the ultimate solution as lots of limitations involve concerning enzyme stability, poor enzyme extraction and higher purification expenditures. In this context, algal enzyme systems could be potential way out. Algae are most ancient microbial entities which exhibit in diverse forms of shape, size, metabolic activities which make it industrially sustainable cellular factories for value added enzymes reservoirs in diverse environmental conditions. Based on this fact, current chapter deals with extensive summarization on productions of algal enzymes as industrially feasible and sustainable way forward.

Keywords: *Algal enzymes, industrial microbiology, bioprocess engineering, enzyme production, biorefinery approach, sustainable bioprocess.*

10.1. Introduction

Microbial enzymes are biomolecules (mostly protein) that acts as biocatalysts in diverse ranges of microbial communities. Microbial enzymes regulate the rates of the biochemical reactions by converting cheaper feedstocks i.e., substrate to value added products through activation energy minimization. At present, microbial enzymes from bacteria and fungi are extensively being used in food, agriculture, cosmetic, pharmaceutical industries concerning its high specificity, catalytic efficiency and minimal by product generation in large scale industrial production. These microbial enzymes include cellulase, amylase, protease, lipase, glucose oxidase, pectinase and tannase. However, poor enzyme extraction, enzyme stability and higher purification expenditure of these fungal and bacterial enzymes have limited their use in industrial applications. Fungal and bacterial enzymes cause lots of ill impacts on human health and environment (Tokuoka et al, 2010). Industrial valuable enzyme streptokinase of *Streptococcus sp* causes acute allergic reaction. Moreover, enzyme carbapenemase in *E. coli* lacks of sensitivity towards lower level of carbapenem in its host range (Lolans et al., 2010). Fungal lipases in *Aspergillus sp., Rhizopus sp.* have been causing cramping, nausea, diarrhea symptoms. *E. coli, Bacillus subtilis* bacteria and fungi like *Trichoderma reesei* produce the enzyme cellulase whose usage is restricted due to higher purification cost toward industrial sustainability. In this context enzymes derived from algae can be a major solution to this problem. However, it has been reported that current global enzymatic marketplace is dominated by fungal and bacterial enzymes. But world microalgae sales are predicted to overcome 75 million USD by 2026 along with global algal biomass market is about 3.8–5.4 billion USD and annual production of ~7000 tonnes (Spier et al., 2020).

Algae are the most ancient microorganism. They are chlorophyll containing microorganisms ranging from microscopic unicellular (*Chlorella)* to large multicellular (giant kelp) they are polyphyletic and mostly photoautotrophic (John, 1986). Algae are photo-synthetically more efficient than terrestrial plants (John, 1986) with higher biomass productivities, faster growth rates, and the highest CO_2 fixation and O_2 production rates when compared with higher plants. Algae can effectively remove the varieties of pollutants and nutrients from the natural water bodies (Gudisa, 2019). They are found in a variety of habitats such as fresh water, sea water, snow, soil or according to Algae base (www.algaebase.org) 162,628 species of algae have been estimated having huge potential to biosynthesize high value-added biomolecules including microbial enzymes (Gupta et al., 2010; Piotrowski, 2008; Katiyar et al., 2017; Bhatia et al., 2018; Wells et al., 2017). Algae have several enzymes mediated metabolic pathways that utilizes enzymes i.e., phosphatase, nitrogenase, carbonic anhydrase, hydrogenase etc. (Fig. 10.1) towards its energy and biomass generations which makes algae have become industrially sustainable reservoirs for algal enzyme production (Sakthinathan, 2018). *Cholophyta, Rhodophyta* including *Saccharina japonica* produce C5 epimerase or alginate lyase, SOD and catalase which provides metal stress tolerance during biorefinery approach towards algal enzyme productions. *Botryococcus braunii* generates synthetase which produces nutritional precursors i.e., bile acid, lipoprotein, steroid (Babu et al., 2014) which accelerates the production of more algal biomass and enzymes. Algal enzyme biosynthesis has been regulated under the control of diverse ranges of stress factors i.e., molar proportion of N_2, CO_2, phosphorous sulphur including P^H, salinity, temperature, light intensity etc. Algal enzyme production has been directly being associated with algal energy metabolism, biomass generation and rate of CO_2 sequestration (Chu, 2017; Uggetti et al., 2018; Almarashi et al., 2020; Li et al., 2012; Chu, 2017; Kumar & Bera, 2020; Ghosh et al., 2020a; Dall'Osto et al., 2019; Singh & Singh, 2015; D'Mello & Chemburkar, 2018; Yang et al., 2017; Zhang et al., 2020; Subhadra & Edwards, 2011). Therefore, such stress associated responses in algal enzyme production is very important towards the establishment of algal enzyme production as biorefinery approach. However, a few attempts had been taken for enzyme generation by using algal machinery. But it is too scanty. Based on this fact, current extensive literature survey deals with detailed summarization on production of algal enzymes and its associated scenarios as industrially feasible and sustainable way forward.

10.2. Methods of Algal Enzymes Desiccation in Biorefinery Approaches

Amelioration of industrially viable algal enzymes production has been highly associated with higher biomass production following several biorefinery approaches. There are diverse ranges of biorefinery steps have been followed likely centrifugation, flocculation and membrane filtration techniques to attain concentrated form of higher algal biomass accumulation towards algal enzyme extractions (Trivedi et al., 2015; Pragya et al., 2013). However, the impact of the algal biorefinery on environment needs to be assessed concerning water, land and carbon footprints (Ubando et al., 2012). Algal enzymes are being produced in large scale industries for different aspects for the productions of natural food colorants in food industries including cosmetics industries, agricultural, environment and health industrial sectors. In biorefinery, industrially sustainable algae derived enzymes have usually been extracted through various approaches i.e., using buffer, extrusion with French pressure and also ultrasonic sound to disrupt the cell wall of algae. Different buffers are used for extraction of several algal enzymes. The Davison buffer has been used to desiccate photosynthetic enzymes and kinases from brown algae. Gerard & Driscoll buffer has been applied to extract algal ribulose 1,5-bisphosphate carboxylase oxygenase enzyme. The phosphate buffer has been used for closing off of the cyanobacterial lipoxygenase enzymes (Hernández et al., 2017). The French pressure has also been applied for decompression and subsequently rupture of algal biomass without usage of denaturing agents (Shechter, 2016; Corrêa et al., 2021). French pressure tool has been used for the isolation of algae-based enzymes i.e., acetyl-CoA synthetase, acetate kinase, phospho transacetylase, citrate synthase, isocitrate dehydrogenase, isocytrate lyase, urease, urea amidolyase, malate synthase, NADP linked glyceraldehyde-3-P dehydrogenase, ribulose-5-P kinase, NADP linked malic dehydrogenase, sedoheptulose 1,7-diphosphatase, fructose 1,6-diphosphate, glucose-6-P dehydrogenase and P fructokinase (Hernández et al., 2017). The ultrasonication method for cell disruption is based on liquid shear forces. The large-scale ultrasound extraction reactors used in food and chemical industry (Corrêa et al., 2021). Ultrasonic disruption can break down the algal cells, and did not affect enzyme activities; this method was used for the isolation of iso-citric acid dehydrogenase, malic acid dehydrogenase, α-ketoglutaric acid dehydrogenase and succinic acid dehydrogenase from blue-green algae. Immunoprecipitation can also be implemented for algal enzyme extraction (Hernández et al., 2017).

10.3. Algae Derived Enzymes Generations for Sustainable Industrial Application

10.3.1. Algal Carbonic Anhydrase, Cellulase, Amylase, Proteases and Lipases, Glucose oxidase, Pectinase, Tannase

Algal carbonic anhydrase is metalloenzyme with Zn^{2+} at the active site and ubiquitous in vertebrates, eubacteria, algae and cytoplasm of green plant including different clades i.e., α-CAs, β-CAs, γ-CAs and δ- and ζ-CAs. Algal carbonic anhydrase involves in several metabolic processes such as acid–base regulation, biomedical applications (artificial lungs, biosensors and CO_2 sequestration systems, drugs) and biofuel production and transportation of carbon dioxide and bicarbonate. Diverse algal regimes i.e., *Acutodesmus sp., Kirchneriella* sp. and *Desmodesmus sp.* have grown in low CO_2 levels (0.03%) and biosynthesized, carbonic anhydrase enzyme to generate higher biomass through CO_2 sequestration for value added products generations including biofuels (Swarnalatha et al., 2015; Huang et al., 2017; Spier et al., 2020). The cellulases are the major group of enzymes to decompose the cellulose exist in lignocellulosic biomass as wastes. The main function of these enzymes is to disintegrate the polysaccharides units of cellulose to simpler forms i.e., β-glucose towards value added biomolecules in bioenergy sector, food industries, textile industries, detergent industries and agricultural sectors (Arora et al., 2019). There are different families of cellulases includes endoglucanases (EC 3.2.1.4), exoglucanases (EC 3.2.1.91) and β-glucosidase (EC 3.2.1.21) and endocellulase. However, the manufacturing expenditure of cellulose-based industries (based on bacterial and fungal communities) can be minimized if cellulases are extracted from the algal biomass as algae can easily been grown predominantly in photoautotrophic mode.

Chlamydomonas sp., Scenedesmus sp., Dunaliella sp., is the most potential reservoir of the enzyme cellulases. Algae are a polyphyletic group of microorganisms that are adapted in wide range of environments. Their autotrophic mode of nutrition and morphologically diverse nature make them attractive candidates towards cellulases productions (Arora et al., 2019). Amylase is another important enzyme which have great industrial impact and potential (Mogharabi and Faramarzi, 2016;

Spier et al., 2020) which hydrolyses starch, oligosaccharides and polysaccharides. Different classes of amylases include α-amylases (EC3.2.1.1, 1, 4-α-D-glucan glucanohydrolase), maltogenic amylase (EC 3.2.1.33), β-amylase (EC 3.2.1.2), amyloglucosidase (also named glucoamylase, EC 3.2.1.3), pullulanase (EC 3.2.1.41), and isoamylase (EC 3.2.1.68).

Currently commercially available bacteria and fungi-based enzymes downstream processing are costly which increases the overall manufacturing costs. Since last few decades a significant development in algal genomics which encourage that algal machinery might be a good source of this class of enzymes. It would be a low-cost potential host for the production of enzymes and other value-added products (Radakovits et al., 2010; Simair et al., 2017; Souza and Magalha˜es, 2010). At present advanced biorefinery approaches minimize the production cost of algal enzymes preparation and strengthen the economic sustainability of the process (Spier et al., 2020). It has been observed that amylase activity in *Spirogyra sp* is 0.042 μmol.min^{-1}.ml^{-1}, *Oedogonium* sp. is 0.0355 μmol.min^{-1}ml^{-1}, *Chlorella sorokiniana* is 0.027 μmol.min^{-1} ml^{-1} and *Scenedesmus obliquus* is 0.009 μmol.min^{-1}.ml^{-1}. *Chlamydomonas reinhardtii* shows amylase activity of 7.3-42 μmol.h^{-1}(10^{9} cells)$^{-1}$ inside chloroplasts. A large number of terrestrial microalgae such as *Rhizoclonium* sp, *Spirogyra* sp, *Klebsormidium* sp, *Rhizoclonium hieroglyphicum*, and *Oedogonium* sp have also been evaluated and showed amylase activity though the catalytic efficiencies need to ameliorate further towards industrial feasibility (Spier et al., 2020). Another potential industrially value-added enzyme is protease (endo and exo peptidases) which hydrolyse amide linkages in protein biomolecules having application in food, chemicals, and pharmaceutical industries (Aguilar and Sato, 2018; Spier et al., 2020; Kaiser et al., 2017). *Chlorella vulgaris* and *Anabaena variabilis* are the primary algal protease producers having trypsin like catalytic efficiencies though complete characterization of these algal proteases are not yet done (Lockau et al., 1988).

Lipases are known as triacylglycerol acylhydrolases (EC 3.1.1.3) responsible for the hydrolysis of triacylglycerols having long chains having great value in several industries namely food, detergents, pharmaceuticals, and bioremediation, and others (Guerrand, 2017). Recently extracellular lipase activity has been observed in *Botryococcus sudeticus* and microalga *Isochrysis galbana* and *Chlamydomonas reinharditii* (Hubert et al., 2017; Li et al., 2012). However, extensive research needs to be done to uncover the kinetic parameters of these algal lipase enzymes in aforementioned algal regimes towards industrial feasibility. GOX (β-D-glucose: oxygen oxidoreductase, EC 1.1.3.4) is another potential enzyme having industrial aspects in therapeutics specially to manage type II diabetes is to inhibit the function of α- glucosidase. Brown algae *Ascophyllum nodosum* is a good algal microbial cellular factory of glucose oxidase (Attjioui et al., 2021) production. Algal phylogeny will help to know the closeness among different algal members. By using molecular tools, the glucose oxidase gene of *Ascophyllum nodosum* can be transferred to the closely related algal member for large scale production. So, it demands extensive research on this particular ground. Pectinase is another class of enzyme having application in fruit and vegetable juice clarification, textile, waste water treatment, paper and wine producing industries. Green algal (*Ulva lactuca*) biomass is used as a feed stack for the microbial (*Bacillus licheniformis)* submerged (SmF) fermentation to produce in a large-scale pectinase enzyme (Iqbal et al., 2017). Even though there is no direct evidence of existence of pectinase enzyme in any algal regime but further extensive research needs to be required in genetic and metabolic level (Pervez et al., 2017). Tannase is also known as tannin acylhydrolase (EC 3.1.1.20).) It catalyzes the hydrolysis of ester bonds presents in gallotannins, complex tannins and gallic acid esters (Rodriguez et al., 2009) having impacts on food, feed, chemical and pharmaceutical industries. Only one marine *Cyanobacterium phormidium valderianum* BDU 140441 has been reported as an algal source of these tannase enzymes. However, kinetic parameters of algal tannase need to be further investigated in near future to ameliorate its industrial viability (Chandrasekaran & Beena, 2013).

10.3.2. Impacts of Algae Derived Enzymes Approaches Multifaceted Industrial Application

10.3.2.1. Algal Phosphatase as a Biosensor

Phosphatases (alkaline phosphatase and acid phosphatase) are hydrolytic enzymes that cleave the ester bond between the phosphate group and residue of the organic phosphate to release the inorganic phosphate. Algal species like *Micromonas sp, Caloglossa continua, Elimeria siliculosus, Gelidium sp, Chladophora sp, Chlorella vulgaris,* are predominantly known to produce alkaline phosphatase including heavy metal biosensor properties (Dotaniya et al., 2019). Acid phosphatase expresses in fresh water microalgae *Chladophora glomerata* (El-Shahed et al., 2006). Glucan phosphatase (ChlreSEX4) that dephosphorylates amylopectin has been found in green microalgae *Chlamydomonas reinharditii* (Carrillo et al., 2020). *Caloglossa continua*, marine red algae produce Mannitol-1-phosphate specific phosphatase which helps in mannitol biosynthesis. Mannitol-1-phosphate is converted to mannitol by the enzymatic activity of mannitol-1-phosphatase (Iwamoto et al., 2005) towards sweetener production.

Recently alkaline phosphatase gene for PhoX has been found in the green algae beside the Pho A and Pho D genes. While PhoA uses Mg^2+ and Zn+ as cofactors, Pho X uses Ca^+ as cofactor. Depending on the expression level of PhoD and PhoX, Alkaline phosphatase regulates the release of phosphorus from Dissolved organic phosphorus (DOP) (Lin et al., 2018). Acid phosphatase does not have any cofactor. Though industrial use of acid phosphatase is limited, alkaline phosphatase is widely used in many industries like food processing, pharmaceutical industries and wastewater treatment (Mohsenpour et al., 2020). Ascorbic acid-2 phosphate is used as a substrate to produce ascorbic acid, a source of vitamin C (Chaudhuri et al., 2015). Alkaline phosphatase activity is used for water monitoring and bioremediation. Alkaline phosphatase is used as a biosensor for heavy metal ions like *Chlorella vulgaris* for (Cd^2+, Zn+) (Durrieu et al., 2002), *Scenedesmus quadricauda* for (Zn+, Ni+, Cd^2+) (Awasthi, 2012) and *Arthrospira platensis* for (Cd^2+, Hg^{2+}) (Tekaya et al., 2013). Alkaline phosphatase is also used for the precipitation of heavy metal ions (Ghosh et al., 2020b). Alkaline phosphatase enzyme inhibition of *Chlamydomonas reinhardtii* is used for detection of metal ion toxicity of water (Islam et al., 2017; Ma et al., 2019).

10.3.2.2. Algal Nitrilase as Potential Biocatalyst

Nitrilase are sub group of carbon- nitrogen (C-N) hydrolase super family. This enzyme hydrolyses the nitriles (R-CN) and denitriles to form carboxylic acid (R-COOH) and ammonia (NH_3). Nitriles are used as solvents in many industries including plastic, synthetic rubber, textile and pharmaceutical industry (Gupta et al., 2010). Many plants, fungi and algae produce nitrilase although the functions are different (Piotrowski, 2008). Nitrilase are found in many algal species like *Chlorella variabilis, Micromonas sp., Micromonas pusilla, Chlorella pyrenoidosa, Tetraselmis suecica* and *Zobellia galactanivorans* (Lauritano et al., 2019). Algal nitrilase has 4 layers of α-β-β-α sandwich folds containing 289 amino acids. The algal nitrilase contains the conserved catalytic triads (E53, K135, and C169) (Zhang et al., 2014). Although *nit1, nit2, nit3, nit4* genes are found in plant nitrilase, only *nit1* gene is found in the algal nitrilase1 (Günther et al., 2018). Nitrilase are commercially used for degradation of toxic cyanide and nitriles (enzymatic bioremediation) (Martínková et al., 2017) earning the name ' green catalyst' isonicotinic acid is used to treat tuberculosis and HIV infection. *Zobellia galactanivorans* nitrilase is used to convert 4 cyanopyridine to achieve yields of 1.79M isonicotinic acid within 3h via fed batch substrate addition (Sunder et al., 2020). Moreover, *Chlorella* sp., and *S. obliquus* can detoxify 50 to 100 mg/L free cyanide over 25hrs (Gurbuz et al., 2004). Even algal nitrilase has been used to produce iminodiacetic acid which is a precursor of chelating agent and surfactant manufacturing (Shen et al., 2021).

10.3.2.3. TAG Synthesis by DGAT) and Galactosidase in Dairy Industry

Acyl-CoA: diacylglycerols acyltransferase (DGAT) is a key enzyme that catalyzes the formation of triglycerides (precursor for biodiesel production) from diacylglycerols and Acyl-CoA (Chen et al., 2015). Diacylglycerols acyltransferase enzymes are found in various types of algae like *Myrmecia incisa, Lobosphaera incisa, Chlorella ellipsoidea, Chromochloris zofingiensis, Chlorella vulgaris* and *Myrmecia incisa. Vernonia* sp. algal diacylglycerols acyltransferase can increase renewable lipid and/or oil production in biofuel industry. Diacylglycerols acyltransferase (DGAT) the only limited enzyme in the synthesis of triacylglycerol (TAG) is regarded as an important therapeutic target for human obesity and other metabolic syndromes in pharmaceutical industry (Borowitzka, 1995). Galactosidases are an enzyme, which catalyzes the hydrolysis of galactosidase into monosaccharides. Galactosidases are classified in either alpha or beta. Galactosidases are found in plant, algae and animal organs and in various microorganisms (Rosenberg, 2006). α-galactosidase has been isolated in the cell-free extracts of red algae *Gracilaria sordida* and *Gracilaria tenuistipitata* is responsible for the degradation of floridosides (Yu and Pedersen, 1990). The α-galactosidase is also being found in *Nannochloropsis gaditana*. The β-galactosidase is found in different microalgae *Cladophora glomerata, Chlorella variabilis, Chlorella sorokiniana*, a red algae *Tetradesmus obliquus* (formerly known as *Scenedesmus obliquus*). A green algae *Ostreococcus lucimarinus* synthesizes β-galactosidase for lactose hydrolysis. Lactose is converted into glucose and galactose, resulting in a necessary internalization of the glucose and galactose molecules during energy generation (Zanette, et al, 2019). The main industrial application of β-galactosidase is hydrolysis of lactose in milk and dairy product, removal of lactose from milk for lactose intolerant people. This enzyme also widely used in food industry to improve sweetness, flavour, solubility and digestibility of dairy products. *Tetradesmus obliquus* can be used as an alternative of enzyme to produce GOS as high-value added ingredient in whey permeate than Bold's Basal Media (BBM) (Weignerová et al., 2009). The α-d-galactosidase is widely use in sugar industry, paper industry and medicine (Suwal et al., 2019) and advantageous in dairy industry (Bentahar et al., 2019).

10.3.2.4. Glucosidase in Flavour Synthesis

Glucosidase is another class of potential algal enzyme which breaks the glycosidic bonds in oligosaccharide or glycoconjugates towards flavouring biomolecule biosynthesis. The α-glucosidase (gene Gasu_00130 and (EC 3.2.1.20) has been isolated from algal species *Gladieria slulphuraria.* The β-glucosidase (EC 3.2.1.21) of *Monoraphidium neglectum* is encoded by gene MNEG_9835. On the other hand, β-glucosidase of *N. oceanica* is encoded by gene g6403; gene encodes protein containing 556 amino acids (Dong et al., 2021). One of the effective managements is inhibition of glucose absorption is the inhibiting the enzyme α-glucosidase. An edible marine algae *Ecklonia cava produces* α-glucoside inhibitor, which is use in for inhibiting α-glucosidase activity (Lee et al., 2009).

10.3.2.5. Algal Hydrogenase in Bioenergy Production

Hydrogenase is a class of enzyme that catalyze the interconversion of hydrogen to protons and electrons. Structurally hydrogenases are divided into three major groups on the metal components of their active sites, i.e., Ni-Fe (subclass Ni-Fe-Se), Fe, and Fe-Fe hydrogenase: Ni-Fe hydrogenase are composed of two subunits and contain the [Ni-Fe] active centre, three iron–sulfur clusters and a magnesium ion (an iron ion in case of [Ni-Fe-Se] hydrogenase. Ni-Fe-hydrogenase is found in variety of archaea and bacteria, catalyzes hydrogen oxidation and reproduction. Fe-hydrogenase harbours a unique iron–guanylyl pyridinol cofactor (Fe-GP). Fe-hydrogenase is only found in hydrogenotrophic methanogenic archaea that use hydrogen for CO_2 reduction to methane. Fe-Fe-hydrogenase is found in few prokaryotes and green algae, catalyzes hydrogen production (Constant & Hallenbeck, 2013). Algal hydrogenase belongs to the Fe-Fe-hydrogenase family, which is not phylogenetically related to the two other hydrogenase families, the Fe-hydrogenase and Ni-Fe-hydrogenase (Lubitz et al., 2007). High degree of variation is seen in the prokaryotic Fe-Fe hydrogenase in their additional N- and C-terminal iron sulphur (Fe-S) cluster binding domains (F-cluster domains), which is absent in most algal hydrogenase. Generally, most algal hydrogenase homologs are consisting only of the active site domain which is H-cluster domain. Their active site is unique as it comprises a standard [4Fe-4S]-cluster with a cysteine linkage to an unusual 2Fe-subcluster with three CO and two CN ligand and an aza-dithiolato-bridge cofactor is an anion (Engelbrecht & Happe, 2018). Even though they lack additional F-clusters, which mediate electron transfer in prokaryotic Fe-Fe hydrogenase (Artz et al., 2017) small algal enzymes achieve very high turnover rates (Peters et al., 1998). Hydrogenase is discovered in green algae nearly 80 years ago Some species of green algae that are been object of research in the biohydrogen productions field are *Chlamydomonas reinharditii, Chlorella fusca, Scenedesmus obliquus, Ulva lactuca, Chlamydomonas moewusii.* Red algae like *Porphyra umbilicalis, Porphyridium cruentum,* and brown algae *Ascophyllum nodosum* also shows hydrogenase activity (Weaver et al., 1980). Under light driven anaerobic conditions unicellular wild type green algae *Chlamydomonas reinhardtii* has evolved the additional ability to reduce protons to produce hydrogen (H_2) via chloroplast Fe-Fe hydrogenase genes: *hydA1* and *hydA2* (Forestier et al., 2003). Fe-Fe hydrogenase, catalyse the reversible reduction of protons to molecular hydrogen has coupled to the photosynthetic electron transport chain via ferredoxin PETF. Some algal hydrogenases have higher specific activity (Kamp et al., 2008) or different redox partner interaction mechanisms (Engelbrecht et al., 2017). Scientific effort was made to find green algae containing hydrogenase, producing hydrogen is used for human needs. Due their high catalytic efficiency, the biotechnological application of algal Fe-Fe-hydrogenases to produce biohydrogen can be used as clean energy which is abundant and renewable (Bolatkhan et al., 2019). Algal Hydrogenase can be used as a better alternative to the industrial production of hydrogen based on fossil fuels or electricity, which is much energy intensive (Engelbrecht et al., 2017). Algal hydrogenase is a better alternative than chemical method for producing hydrogen and bioelectricity, which is used as a major intermediate for fertilizer production and chemical synthesis (Benjamin & Shaw, 2020; Kim et al., 2011; Varjani et al., 2020). Hydrogenase can be used as an alternative to pt-based catalysts for biological hydrogen production and in technological application, e.g., (Photo-) electrolysers and fuel cell. The Highly active Fe-Fe-hydrogenase have also been successfully used in hydrogen oxidizing fuel cell (Khanna et al., 2017).

10.3.2.6. Carbonic Anhydrase for Carbon Concentrating Mechanism

Carbonic anhydrase (CA) is a ubiquitous enzyme. It is a metalloenzyme contains zinc ion (Zn^{2+}) in its active site. It plays a fundamental role in CO_2 concentrating mechanisms important for uptake of inorganic carbon source (Bhat et al., 2017). The occurrence of carbonic anhydrase in plants confirms by Bradfield in 1947 (Moroney et al., 2001). Carbonic anhydrase divides into six distinct families (alpha, beta, gamma, delta and zeta) shares no significant sequences similarity with each other (Jeon et al., 2016) including alpha-CA, beta-CA, and gamma-CA in C. *reinharditti* while Delta-CA predominates in diatoms. In contrary zeta-CA predominantly has been found in exclusively in bacteria in a few chemolithotrops and marine cyanobacteria. Active site of the carbonic anhydrase contains a zinc ion and shares the electron-rich hydroxide ion to attack the CO_2 to generate bicarbonate molecule (Bhat et al., 2017). Regarding CO_2 assimilation the carbon concentrating mechanism in *C. reinhardtii* the gene expression level of CAs varies considerably. Carbonic anhydrase enzyme founds in various types of algae like *Chlamydomonas reinhardtii*, *Dunaliella salina* and *Dunaliella salina.* Activity of these algal carbonic anhydrase is significantly lower in patients with renal cell carcinoma than in controls. The marker carbonic anhydrase might be potentially important as an additional biochemical tool for diagnosing renal carcinoma. Discovery and development of carbonic anhydrase inhibitors is crucial for their clinical use as antiepileptic, diurectic and antiglaucoma agents (Polle et al., 2009).

10.3.2.7. Algal L-asparaginase in treat Acute Lymphocytic Leukaemia (ALL) Chemotherapy and use of Algal superoxide dismutase, Algal catalase

L-asparaginase is an enzyme uses in acute lymphoblastic leukaemia chemotherapy and acute childhood lymphoblastic leukaemia (Qeshmi et al., 2018). This amidohydrolase enzyme catalyzes the hydrolysis of amino acid asparagine to aspartic acid and ammonia (Ebrahiminezhad et al., 2014). L-asparagine is a non-essential amino acid synthesizes in ordinary human cells by L-asparagine synthetase from aspartic accuses (Qeshmi et al., 2018). Microalgae like *Chlamydomonas spp., Chlorella vulgaris* etc. produces this enzyme (Lauritano et al., 2019). In a DOE approach *Chlorella vulgaris* biosynthesizes 2.61% (v/v) of L-asparaginase having specific enzyme activity of 26.39% (Ebrahiminezhad et al., 2014). Superoxide dismutase [SOD] is an enzyme that converts superoxide radical to normal molecular oxygen O_2 and hydrogen peroxide (H_2O_2) SODs are classified based on the type of metal bound in the active space: Cu / Zn-SOD (Cu / Zn is a cofactor to the superoxide dismutase enzyme), Mn-SOD, Fe-SOD, and Ni-SOD (Zelko et al., 2002). Green algae i.e., *Closterium ehnebergii, Meneghini sp.;* and few species of red algae i.e., *Kappaphycus alcarezi*i and *Carallina pilulifera* biosynthesize SOD enzyme (Ichimura & Kasai, 1984). Microalgae are exposed to various environmental stresses including variable salinity, heavy metals, endocrine disruptive chemicals (EDCs) and other toxic pollutants and as a result of such stresses and reactions they form reactive oxygen species or ROS (Yoboue et al., 2014).

The levels of ROS and the types of protective molecules are related to the stress levels in the cells, and thus their toxicity has been investigated in a variety of organisms, including microalgae. SOD may be an effective antioxidant therapeutic agent for managing the detrimental consequences of inflammatory diseases, as well as for mitigating other conditions associated with uncontrolled overproduction of superoxide (Levin, 2005). SOD used in cosmetic industry as well (Wang et al., 2015). When SOD level increased in plants, can protect the host against physical and chemical stress, and improve the biomass production with larger shoot, crown, and root system (Sahoo et al., 2001; Kim et al., 2011). The SOD of dietary antioxidant bioavailability depends on the stability of the food matrix in controlling food processing, food deprivation, antioxidant stability, inactivation of lipophilic antioxidants, especially carotenoids and isometric forms (Ratnam et al., 2006). Therefore, it could easily be speculated that amelioration of Superoxide dismutase [SOD] biosynthesis in algal regimes could be potential way out towards biorefinery approach for human health care.

Catalase is a common enzyme found in almost all organisms including algae. Few species of algae like *Shewanella xiamenesis*, green algae like *Cyanophyta anabaena* produce this enzyme (Okuno et al., 2008, Luo & Rando, 2003). Catalase enzymes have been employed in various aspects of bioremediation where it is known to act as an indicator of the hydrocarbon degradation of crude oil contaminated soils, cheese production, baking industry, bleaching agent of textile industries, diagnosing agent for detection of renal cell carcinoma, plant root protector from pathogens (Ng et al., 2014; Geciova et al., 2002; Grigoras, 2017; Chico et al., 2013; Pirinççi et al., 2017; Anderson et al., 2001). However, very little information has been available on kinetic properties of these algal catalase enzymes towards its large-scale production in near future.

10.3.2.8. A Ubiquitous Enzyme Lipoxygenase Biosynthesizes in Algae

Lipoxygenases are a family of iron (Fe^{2+}) containing enzymes (LOX gene) most of which catalyzes the dioxygenation of polyunsaturated fatty acids in lipids containing a cis-1,4-pentadiene into cell signaling agents that serve diverse role as anticrime signals. The red algae *Pyropia hyatenesis* Kikuchi and Myata have a LOX with multiple catalytic sites in the knowledge that oxylipins are involved in the protective mechanism of algae (Brash, 1999; Andreou et al., 2009; Xie et al., 2013). Lipoxygenases have ability to engineer cropping plants to enhance the resistant against the biotic or abiotic stress factor (Hu et al., 2013). Moreover, Lipoxygenases is widely used in food industry to improve aroma, baking properties of food (Lu et al., 2013; Patel et al., 2015). Lipoxygenase use for induction of 15-LOX-1-mediated down regulation of a PPAR-gamma and COX-2 pathway by honokiol is a therapeutic strategy for gastric cancer (Liu et al., 2010).

10.3.2.9. Algal thiolase in Biofuel Production

Thiolases is a widely distributed enzyme. It catalyzes the Claisen condensation of two acetyl-CoA to generate acetoacetyl-CoA (Hernández et al., 2017; Price et al., 2020). Thiolase is extensively distributed in prokaryotes, eukaryotes as well as in algae. Some algae which can produce thiolase are *Mougeotia sp*, *Bumilleriopsis sp*, *Euglena gracilis* and *Eremosphaera sp*. etc. Thiolase is located on both peroxisomes and mitochondria in Euglena *gracilis Klebs* and *Eremosphaera* (Hernández et al., 2017). Thiolase enzyme has great impact for PHD production in blue green algae or cyanobacteria depending on *phaA and phaB* genes expression in two separate operons (Price et al., 2020; Bharti and Swetha, 2016). Thiolase are the key enzyme in the biosynthesis of sterols by mevalonic acid pathway and carotenoids biosynthesis in algae including *Cyanophora sp., Galdieria sp., Cyanidium sp., Mesostigma sp.* and *Euglena sp.* Even algal thiolase does play important applications towards bioplastics PHB or poly-hydroxybuterate, biofuel (biobutanol, sterol and isoprenoids), biopigment (carotenoids) productions (Hernández et al., 2017).

10.3.2.10. Algal ascorbate peroxidase as an Antioxidative Enzyme and Phytoene synthase-important Enzyme in Carotenoids Production

Ascorbate peroxidase is used in bioremediation of polluted water by oxidation. It also used in paper industry as delignification. It can be used in polymerization; in hydrogen peroxide quantification. It also can act as a bio sensor (Singh et al., 2019). For the survival under stress condition Ascorbate peroxidase is produced in different microbes including algae (Patel & Bhatt, 2020). Ascorbate peroxidase converts H_2O_2 to H_2O through ascorbate as an electron donor (Rezayian et al., 2019). Ascorbate peroxidase has been biosynthesized in algal regime i.e. *Spirodela polyrhiza, Chlorella pyrenoidosa, Prototheca moriformis, Ochromonas danica etc.* Algal phytoene synthase (*psy* gene) catalyzes the first committed step in carotenoids biosynthesis and its corresponding regulation in presence of NADPH as a cofactor (Tian, 2015; Saini et al., 2020). Carotenoids (β-carotenoids, lutein, and astaxanthin) are most commercial biopigments because of its health and industrial application. Some algae produce carotenoids which include *Haematococcus pluvialis, Chlorella zofingiensis, Chlamydomonas reinhardtii, Volvox carteri*, and *Chlorella vulgaris*. It has been reported that under stress and nitrogen starvation conditions *psy* gene has been up-regulated in *chlorella zofingiensis.* It results in almost four folds increase in astaxanthin production. Moreover, higher expression of phytoene synthase is used to synthesize carotenoids like beta carotene, astaxanthin, lutein etc (Saini et al., 2020). Beta carotene has anti oxidant property. It is the precursor of vitamin A. It prevents muscular degeneration, asthma. Thus, it is used in pharmaceutical and cosmetics industry. Astaxanthin is an antioxidant. It protects from UV rays. It uses as food colorants. It enhances immune functions. It has anti aging property. It is used in pharmaceutical, neutraceuticals and cosmetics industry. Lutein is a feed additive, food colorant. It helps to regulate cardiovascular diseases, cancers, cognitive function, and age-related macular degeneration (AMD) in human. It is used in pharmaceutical industry. It is also used as a food colorant, and feed additive (Saini et al., 2020).

10.3.2.11. Industrial Aspect of Algal nitrogenase Enzyme

This enzyme is essential for atmospheric nitrogen fixation. Nitrogenase is an enzymatic complex consists of component I (dinitrogenase) is a Mo-Fe protein also contain S. It is tetramer, consisting of two α and two β subunits of 240 KDa. Component II is an Fe-S protein, known as dinitrogenase reductase, it is a dimer. Mainly *nif* genes of nitrogenase gene clusters are responsible for nitrogen fixation. Nitrogenase enzyme isolated from the algal specie like *Gloethece sp, Plectonema sp, Pseudoanabaena sp, Anabaena sp, Calothrix sp, Nodularia sp, Nostoc sp, Stigonema sp, Trichodesmium* sp. Some cyanobacteria can form specialiszed cell (heterocysts), after receiving some environmental signals like, when nitrogen sources are limited, some cells from thallus change into heterocysts (Hernandez et al., 2017; Mishra et al., 2019). In cyanobacteria, mature heterocyst is the actual site of nitrogen fixation. It possesses oxygen-sensitive nitrogenase enzyme complex (nifHDK) which reduces atmospheric dinitrogen to ammonia (NH_3). In the presence of nitrogen sources other than ammonium, or under nitrogen starvation, NtcA activates the transcriptional regulator of oxygen-sensitive nitrogenase enzyme complex (nifHDK) associated gene clusters (Esteves-Ferreira et al., 2017). In non heterocystous cyanobacteria, nitrogen fixation is temporally separated from oxygen derived photosynthesis. Where photosynthetic O_2 development takes place during daytime while atmospheric nitrogen fixed during dark (Mishra et al., 2019; Bhatia and Paliwal, 2010). In *Synechococcus, Ntc*A transcripts accumulated during the day, indicating an insufficiency of fixed N_2 and promoting an accumulation of *nif* transcripts during the evening, when the net oxygen evolution is low or negative (Esteves-Ferreira et al., 2017). The nitrogenase enzyme can fix atmospheric nitrogen that is very useful for improving agricultural crop productivities. This nitrogenase enzyme can also be used to formulate organic fertilizers in association with whole cell nitrogen fixing algae communities. The nitrogenase complex has also a high potential for biofuel production (Abdelaziz et al., 2014) under nitrogen limiting conditions.

10.3.2.12. Putative Algal chromium reductase Enzyme carries out Bioremediation and Algal lypolitic acid hydrolases 1 Enzyme in Health Application

Putative chromium reductase is one of the enzymes that helps in bioremediation of waste water specially Chromium (VI) ion contamination in industrial waste effluents. Putative chromium reductase enzyme does exist in different algal regimes i.e., Chlorella vulgaris. Putative chromium reductase of Chlorella vulgaris detoxifies the highly toxic chromium (VI) to less toxic chromium (III) ion (Vingiani et al., 2019; Cervantes et al., 2001; Yen et al., 2017). However, Putative chromium reductase research does exist in still its infancy; extensive research needs to be carried out in near future. Lypolitic acid hydrolases 1 (LAH1) is derived from algal regimes i.e., *Pseudonitzchia multistriata and Pseudonitzchia arenysensis. Pseudonitzchia multistriata produces lypolitic acid hydrolases 1 enzyme engaged in the synthesis of oxylipins, which are the secondary metabolites of Pseudonitzchia multistriata.* Lypolitic acid hydrolases 1 is involved in the release of the fatty acid precursors of oxylipins and tested their galacto-lipase activity in vitro (Vingiani et al., 2019). As per biorefinery aspects, algal LAH1 enzymes of P. multistriata and P. arenysensis may play the same role in the production of oxylipins and development of a chemical mechanism of defense against diatom. Algal LAH1 enzymes are used in health industrial sectors. Moreover, Algal LAH1 enzymes are involved in the oxylipins biosynthesis; it depicts anti-cancer activity and anti-predator activity (Adelfi et al., 2019).

10.4. Conclusion

Microalgae are sustainable and ecofriendly source of biological products. Algae can produce different bio-compounds like pigments, toxins and enzymes that can be considered as high value biomolecules. In this chapter, an extensive survey of literature has been summarized on different algal educed enzymes and their applications. These enzymes exhibit interesting properties for industry like pharmaceutical, cosmetics, food, bioremediation and bio-monitoring. These enzymes represent an opportunity for biotechnological development. It is mainly focused on enzymes involved in pathways for lipid synthesis with applications as neutraceuticals, cosmeceuticals, or as a feedstock for biofuel i.e., biohydrogen and biodiesel production. The advantage of using algae for enzyme production as it can be grown without carbon or nitrogen sources, and can consume small quantities of toxic intermediate and/or end metabolites. The future approaches to maximize the algal enzyme production (molar yield and productivities) need to be focused on three possible ways which includes a. usage of ever increasing amounts of Omics dataset for the production of high valuable products in algae through over expression of one or more enzyme by genomic editing.; b. identification and characterization of metabolic pathway for existing and/or novel enzyme production using metabolomics datasets; c. search for novel gene clusters which involve in the production of novel enzymes utilizing transcriptomics data repositories. Therefore, algae educed enzyme could be a potential alterative over bacterial and/or fungal enzymes (Vingiani et al., 2019; Hernández et al., 2017) if productivities and yields of algal enzymes ameliorate via scale up studies as sustainable biorefinery future way-out.

Acknowledgements

Author would like to thank JIS University, Kolkata and JIS Group of Education Initiatives for encouragements. Authors are also very grateful to Miss. Shrestha Debnath (Ph.D. Scholar of Dr. Ghosh's MEAB laboratory, JIS University Kolkata) for her immense support and help during this book chapter preparation.

Figure Legends

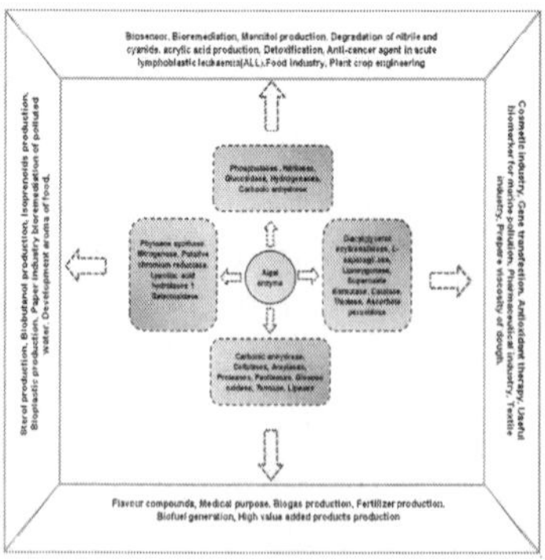

Figure 10.1. - Algae educed enzyme and their industrial application

Multiple algal regimes (e.g. *Ascophyllum nodosum, Plectonema sp, Pseudoanabaena sp, Anabaena sp, Calothrix sp, Nodularia sp, Nostoc sp, Stigonema sp., Pseudonitzschia multistrata, Pseudonitzschia arenysensis, Cyanobacterium phormidium valderianum BDU 140441, Botryococcus sudeticus etc.)* have been depicted towards various commercial purposes. Different genes are also responsible for accelerate the algal enzymes encode activities. Such as *Caloglossa continua* (PhoA), *Chlorella vulgaris* (PhoD), *Chlamydomonas reinhardtii* (PhoX), *Chladophora glomerata (acid), Chlorella variabilis* (NIT1), *Micromonas pusilla, Chlorella pyrenoidosa, Tetraselmis suecica, Gladieria slulphuraria, (α-glucosodase) (*Gasu_00130), *Monoraphidium neglectum, Nannochloropsis oceanica (B-glucosidase)* (MNEG_9835 and g6403), *Green algae Chlamydomonas reinharditii, Chlorella fusca, Scenedesmus obliquus, Ulva lactuca, Chlamydomonas moewusii. Red algae Porphyra umbilicalis , Porphyra cruentum, and brown algae Ascophyllum nodosu* (HydA), *Dunaliella salina, Chlamydomonas reinhardtii* (DsCA1a, DsCA1b, DsCA2a,DsCA2b, DsCA3, DsgCA1,DsgCA2, DsgCA3), *Myrmecia incisa Reisigl, Lobosphaera incisa, Chlorella ellipsoidea, Chromochloris zofingiensis* (DGAT1, DGAT2, DGAT3, MiDGAT1, MiDGAT2A, MiDGAT2B), *Chlamydomonas sp., Chlorella vulgaris* (MCCS 025 18S ribosomal RNA gene, ASPG gene), *Pyropia hyatenesis, Dictyopteris sp.* (LOXs gene), *Closterium ehnebergi, Kappaphycus alcarezii, Carolina pilulifera* (SOD 1 gene for *Closterium ehnebergii,* MmCU/ZnSOD gene for *Kappaphycus alcarezii*), *Shewanella xiamenesis, Cyanophyta anabaena, Shewanella oneidensis* (Kat B gene), *Mougeotia sp, Bumilleriopsis sp, Euglena gracilis Eremosphaera sp.* (phaA gene), *Spirodela polyrhiza, Chlorella pyrenoidosa, Prototheca moriformis, Ochromonas danica, Thalassiosira pseudonana, Volvox carteri f. nagariensis* (Apxs gene for *Thalassiosira pseudonana* ,apx1 gene for *Volvox carteri f. nagariensis*), *Haematococcus pluvialis, Chlorella zofingiensis ,Chlamydomonas reinhardtii* (Psy gene), *Gloethece sp., Plectonema sp., Pseudoanabaena sp., Anabaena sp., Calothrix sp., Nodularia sp., Nostoc sp., Stigonema sp.* (Nif genes), *Chlorella vulgaris* (Chromate reductase gene), *Pseudo-nitzschia multistrata, Pseudo-nitzschia arenysensis* (PaLAH1, PmLAH1), *Gracilaria sordida and Gracilaria tenuistipitata* (α-galactosidase) (GALA lacZ).

References

Abdelaziz, A. E. M., Ghosh, D. & Hallenbeck, P. C. (2014). Characterization of growth and lipid production by Chlorella sp. PCH90, a microalga native to Quebec. Bioresource technology, 156, 20-28.

Adelfi, M. G., Vitale, R. M., d'Ippolito, G., Nuzzo, G., Gallo, C., Amodeo, P. ... & Fontana, A. (2019). Patatin-like lipolytic acyl hydrolases and galactolipid metabolism in marine diatoms of the genus Pseudo-nitzschia. Biochimica et Biophysica Acta (BBA)-Molecular and Cell Biology of Lipids, 1864(2), 181-190.

Aguilar, J. G. & Sato, H. H. (2018). Microbial proteases: production and application in obtaining protein hydrolysates. Food Research International, 103, 253-262.

Almarashi, J. Q., El-Zohary, S. E., Ellabban, M. A. & Abomohra, A. E. F. (2020). Enhancement of lipid production and energy recovery from the green microalga Chlorella vulgaris by inoculum pretreatment with low-dose cold atmospheric pressure plasma (CAPP). Energy Conversion and Management, 204, 112314. zDos Santos

Anderson, A. J. & Miller, C. D. (2001). Catalase activity and the survival of Pseudomonas putida, a root colonizer, upon treatment with peracetic acid. Canadian journal of microbiology, 47(3), 222-228.

Andreou, A. & Feussner, I. (2009). Lipoxygenases–structure and reaction mechanism. Phytochemistry, 70(13-14), 1504-1510.

Andreou, A., Brodhun, F. & Feussner, I. (2009). Biosynthesis of oxylipins in non-mammals. Progress in lipid research, 48(3-4), 148-170.

Arora, P., Shukla, V. K. & Tiwari, A. (2019). Algal Cellulases. In New and Future Developments in Microbial Biotechnology and Bioengineering (pp. 283-295). Elsevier.

Artz, J. H., Mulder, D. W., Ratzloff, M. W., Lubner, C. E., Zadvornyy, O. A., LeVan, A. X. ... & Peters, J. W. (2017). Reduction potentials of [FeFe]-hydrogenase accessory iron–sulfur clusters provide insights into the energetics of proton reduction catalysis. Journal of the American Chemical Society, 139(28), 9544-9550.

Attjioui, M., Ryan, S., Ristic, A. K., Higgins, T., Goñi, O., Gibney, E. R. ... & O'Connell, S. (2021). Comparison of edible brown algae extracts for the inhibition of intestinal carbohydrate digestive enzymes involved in glucose release from the diet. Journal of nutritional science, 10.

Awasthi, M. (2012). Relevance of Alkaline Phosphatase activity of immobilized green algae and cyanobacteria for heavy metal toxicity monitoring. Journal of Material and Environmental Science, 3(3), 446-51.

Babu, M. Y., Palanikumar, L., Nagarani, N., Devi, V. J., Kumar, S. R., Ramakritinan, C. M. & Kumaraguru, A. K. (2014). Cadmium and copper toxicity in three marine macroalgae: evaluation of the biochemical responses and DNA damage. Environmental Science and Pollution Research, 21(16), 9604-9616.

Benjamin, K. R. & Shaw, J. (2020). Energy biotechnology 2020. Current opinion in biotechnology, 62, vii-ix.

Bentahar, J., Doyen, A., Beaulieu, L. & Deschênes, J. S. (2019). Investigation of β-galactosidase production by microalga Tetradesmus obliquus in determined growth conditions. Journal of Applied Phycology, 31(1), 301-308.

Bharti, S. N. & Swetha, G. (2016). Need for bioplastics and role of biopolymer PHB: a short review. J Pet Environ Biotechnol, 7(272), 2.

Bhat, F. A., Ganai, B. A. & Uqab, B. (2017). Carbonic anhydrase: mechanism, structure and importance in higher plants. Asian J Plant Sci Res, 7(3), 17-23.

Bhatia, L. & Paliwal, S. (2010). Banana peel waste as substrate for ethanol production. International Journal of Biotechnology and Bioengineering Research, 1(2), 213–218.

Bhatia, L., Singh, A., Chandel, A. & Singh, O. M. (2018). Biotechnological Advancements in cellulosic ethanol production. O. V. Singh and A. K. Chandel (eds.), Sustainable Biotechnology- Enzymatic Resources of Renewable Energy, https://doi.org/10.1007/978-3-319-95480-6.

Bolatkhan, K., Kossalbayev, B. D., Zayadan, B. K., Tomo, T., Veziroglu, T. N. & Allakhverdiev, S. I. (2019). Hydrogen production from phototrophic microorganisms: reality and perspectives. international journal of hydrogen energy, 44(12), 5799-5811.

Borowitzka, M. A. (1995). Microalgae as sources of pharmaceuticals and other biologically active compounds. Journal of applied phycology, 7(1), 3-15.

Brash, A. R. (1999). Lipoxygenases: occurrence, functions, catalysis, and acquisition of substrate. Journal of Biological Chemistry, 274(34), 23679-23682.

Carrillo, J. B., Torresi, F., Morales, L. L., Ricordi, M., Gomez-Casati, D. F., Busi, M. V. & Martin, M. (2020). Identification and characterization of ChlreSEX4, a novel glucan phosphatase from Chlamydomonas reinhardtii green alga. Archives of biochemistry and biophysics, 680, 108235.

Cervantes, C., Campos-García, J., Devars, S., Gutiérrez-Corona, F., Loza-Tavera, H., Torres-Guzmán, J. C. & Moreno-Sánchez, R. (2001). Interactions of chromium with microorganisms and plants. FEMS microbiology reviews, 25(3), 335-347.

Chandrasekaran, M. & Beena, P. S. (2013). Tannase: source, biocatalytic characteristics, and bioprocesses for production. In Marine enzymes for biocatalysis (pp. 259-293). Woodhead Publishing.

Chaudhuri, G., Venu-Babu, P., Dalal, D. & Thilagaraj, W. R. (2015). Application of alkaline phosphatase for heavy metals precipitation using ascorbic acid 2-phosphate as an effective natural substrate. International journal of environmental science and technology, 12(12), 3877-3886.

Chen, C. X., Sun, Z., Cao, H. S., Fang, F. L., Ouyang, L. L. & Zhou, Z. G. (2015). Identification and characterization of three genes encoding acyl-CoA: diacylglycerol acyltransferase (DGAT) from the microalga Myrmecia incisa Reisigl. Algal research, 12, 280-288.

Chico, D., Aldaya, M. M. & Garrido, A. (2013). A water footprint assessment of a pair of jeans: the influence of agricultural policies on the sustainability of consumer products. Journal of Cleaner Production, 57, 238-248.

Chu, W. L. (2017). Strategies to enhance production of microalgal biomass and lipids for biofuel feedstock. European Journal of Phycology, 52(4), 419-437.

Constant, P. & Hallenbeck, P. C. (2013). Biohydrogen: Chapter 5. Hydrogenase. Elsevier Inc. Chapters.

Corrêa, P. S., Morais Júnior, W. G., Martins, A. A., Caetano, N. S. & Mata, T. M. (2021). Microalgae biomolecules: Extraction, separation and purification methods. Processes, 9(1), 10.

D'Mello, B. R. & Chemburkar, M. S. (2018). Effect of temperature and pH variation on biomass and lipid production of Auxenochlorella pyrenoidosa. Res J Life Sci Bioinform Pharm Chem Sci, 4, 378-387.

Dall'Osto, L., Cazzaniga, S., Guardini, Z., Barera, S., Benedetti, M., Mannino, G. ... & Bassi, R. (2019). Combined resistance to oxidative stress and reduced antenna size enhance light-to-biomass conversion efficiency in Chlorella vulgaris cultures. Biotechnology for biofuels, 12(1), 1-17.

Dong, S., Liu, Y. J., Zhou, H., Xiao, Y., Xu, J., Cui, Q. ... & Feng, Y. (2021). Structural insight into a GH1 β-glucosidase from the oleaginous microalga, Nannochloropsis oceanica. International Journal of Biological Macromolecules, 170, 196-206.

Dotaniya, M. L., Aparna, K., Dotaniya, C. K., Singh, M. & Regar, K. L. (2019). Role of soil enzymes in sustainable crop production. In Enzymes in Food Biotechnology (pp. 569-589). Academic Press.

Durrieu, C. & Tran-Minh, C. (2002). Optical algal biosensor using alkaline phosphatase for determination of heavy metals. Ecotoxicology and Environmental Safety, 51(3), 206-209.

Ebrahiminezhad, A., Rasoul-Amini, S., Ghoshoon, M. B. & Ghasemi, Y. (2014). Chlorella vulgaris, a novel microalgal source for L-asparaginase production. Biocatalysis and Agricultural Biotechnology, 3(2), 214-217.

El-Shahed, A. M. (2006). Isolation and characterization of phosphatase enzyme from the freshwater macroalga Cladophora glomerata Kutzing (chlorophyta). Pakistan J. Biol. Sci., 9, 2456-2461.

Engelbrecht, V. & Happe, T. (2018). [FeFe]-hydrogenases from green algae. Methods in enzymology, 613, 203-230.

Engelbrecht, V., Rodríguez-Maciá, P., Esselborn, J., Sawyer, A., Hemschemeier, A., Rüdiger, O. ... & Happe, T. (2017). The structurally unique photosynthetic Chlorella variabilis NC64A hydrogenase does not interact with plant-type ferredoxins. Biochimica et Biophysica Acta (BBA)-Bioenergetics, 1858(9), 771-778.

Esteves-Ferreira, A. A., Cavalcanti, J. H. F., Vaz, M. G. M. V., Alvarenga, L. V., Nunes-Nesi, A. & Araújo, W. L. (2017). Cyanobacterial nitrogenases: phylogenetic diversity, regulation and functional predictions. Genetics and molecular biology, 40, 261-275.

Forestier, M., King, P., Zhang, L., Posewitz, M., Schwarzer, S., Happe, T. ... & Seibert, M. (2003). Expression of two [Fe]-hydrogenases in Chlamydomonas reinhardtii under anaerobic conditions. European Journal of Biochemistry, 270(13), 2750-2758.

Geciova, J., Bury, D. & Jelen, P. (2002). Methods for disruption of microbial cells for potential use in the dairy industry—a review. International Dairy Journal, 12(6), 541-553.

Ghosh, D., Bera, K., Bera, P., Bhattacharya, M., Samanta, B., Mondal, S. ... & Mondal, P. (2020a) Systems Biology Approach to Screen and Identify Algae-Based Alkaline Phosphatases as Heavy Metal-Detecting Biosensors. In Advances in Bioprocess Engineering and Technology, (pp. 431-437). Springer, Singapore.

Ghosh, D., Debnath, S. & De, K. (2020b). Impact of Stress Factors on Plants for Enhancing Biomass Generations Towards Biofuels. In Plant Stress Biology (pp. 313-328). Apple Academic Press.

Grigoras, A. G. (2017). Catalase immobilization—A review. Biochemical Engineering Journal, 117, 1-20.

Gudisa, I. (2019). Algal Multipurpose Benefit. Journal of Natural Sciences Research, 9, 23.

Guerrand, D. (2017). Lipases industrial applications: focus on food and agroindustries. OCL Oilseeds and fats crops and lipids, 24(4), D403.

Günther, J., Irmisch, S., Lackus, N. D., Reichelt, M., Gershenzon, J. & Köllner, T. G. (2018). The nitrilase PtNIT1 catabolizes herbivore-induced nitriles in Populus trichocarpa. BMC plant biology, 18(1), 1-12.

Gupta, V., Gaind, S., Verma, P. K., Sood, N. & Srivastava, A. K. (2010). Purification and characterization of intracellular nitrilases from Rhodococcus sp.-potential role of periplasmic nitrilase. African Journal of Microbiology Research, 4(11), 1148-1153.

Gurbuz, F., Ciftci, H., Akcil, A. & Karahan, A. G. (2004). Microbial detoxification of cyanide solutions: a new biotechnological approach using algae. Hydrometallurgy, 72(1-2), 167-176.

Hernández, E., Lobato-Benítez, C. & Hernández, C. (2017). Algal enzymes, biotechnological potential uses: A review Enzimas algales, usos biotecnológicos potenciales: Una revisión. Cymbella, 3, 1-15.

Hu, T., Zeng, H., Hu, Z., Qv, X. & Chen, G. (2013). Overexpression of the tomato 13-lipoxygenase gene TomloxD increases generation of endogenous jasmonic acid and resistance to Cladosporium fulvum and high temperature. Plant Molecular Biology Reporter, 31(5), 1141-1149.

Huang, G., Chen, S., Dai, C., Sun, L., Sun, W., Tang, Y. ... & Ma, H. (2017). Effects of ultrasound on microbial growth and enzyme activity. Ultrasonics Sonochemistry, 37, 144-149.

Hubert, J., Nuzillard, J. M. & Renault, J. H. (2017). Dereplication strategies in natural product research: How many tools and methodologies behind the same concept? Phytochemistry Reviews, 16(1), 55-95.

Ichimura, T. & Kasai, F. (1984). Post-zygotic isolation between allopatric mating groups of Closterium ehrenbergii Meneghini (Conjugatophyceae). Phycologia, 23(1), 77-85.

Iqbal, S., Irfan, M., Tabassum, F., Shakir, H. A. & Qazi, J. I. (2017). Application of Box-Behnken design for optimization of different pre-treatments conditions for cellulase production. Journal of Northeast Agricultural University, 24(3), 51-59.

Islam, M. S., Sazawa, K., Hata, N., Sugawara, K. & Kuramitz, H. (2017). Determination of heavy metal toxicity by using a micro-droplet hydrodynamic voltammetry for microalgal bioassay based on alkaline phosphatase. Chemosphere, 188, 337-344.

Iwamoto, K. & Shiraiwa, Y. (2005). Salt-regulated mannitol metabolism in algae. Marine Biotechnology, 7(5), 407-415.

Jeon, H., Jeong, J., Baek, K., McKie-Krisberg, Z., Polle, J. E. & Jin, E. (2016). Identification of the carbonic anhydrases from the unicellular green alga Dunaliella salina strain CCAP 19/18. Algal research, 19, 12-20.

John Pirt, S. (1986). The thermodynamic efficiency (quantum demand) and dynamics of photosynthetic growth. New phytologist, 102(1), 3-37.

Kaiser, D., Kowalski, N. & Waniek, J. J. (2017). Effects of biofouling on the sinking behavior of microplastics. Environmental Research Letters, 12(12), 124003.

Kamp, C., Silakov, A., Winkler, M., Reijerse, E. J., Lubitz, W. & Happe, T. (2008). Isolation and first EPR characterization of the [FeFe]-hydrogenases from green algae. Biochimica et Biophysica Acta (BBA)-Bioenergetics, 1777(5), 410-416.

Katiyar, R., Gurjar, B. R., Biswas, S., Pruthi, V., Kumar, N. & Kumar, P. (2017). Microalgae: an emerging source of energy-based bio-products and a solution for environmental issues. Renewable and Sustainable Energy Reviews, 72, 1083-1093.

Khanna, N., Esmieu, C., Mészáros, L. S., Lindblad, P. & Berggren, G. (2017). In vivo activation of an [FeFe] hydrogenase using synthetic cofactors. Energy & Environmental Science, 10(7), 1563-1567.

Kim, M. S., Kim, D. H., Son, H. N., Ten, L. N. & Lee, J. K. (2011). Enhancing photo-fermentative hydrogen production by Rhodobacter sphaeroides KD131 and its PHB synthase deleted-mutant from acetate and butyrate. international journal of hydrogen energy, 36(21), 13964-13971.

Kumar, A. & Bera, S. (2020). Revisiting nitrogen utilization in algae: a review on the process of regulation and assimilation. Bioresource Technology Reports, 100584.

Lauritano, C., De Luca, D., Amoroso, M., Benfatto, S., Maestri, S., Racioppi, C. ... & Ianora, A. (2019). New molecular insights on the response of the green alga Tetraselmis suecica to nitrogen starvation. Scientific reports, 9(1), 1-12.

Lee, S. H., Karadeniz, F., Kim, M. M. & Kim, S. K. (2009). α-Glucosidase and α-amylase inhibitory activities of phloroglucinal derivatives from edible marine brown alga, Ecklonia cava. Journal of the Science of Food and Agriculture, 89(9), 1552-1558.

Levin, E. D. (2005). Extracellular superoxide dismutase (EC-SOD) quenches free radicals and attenuates age-related cognitive decline: opportunities for novel drug development in aging. Current Alzheimer Research, 2(2), 191-196.

Li, X., Benning, C. & Kuo, M. H. (2012). Rapid triacylglycerol turnover in Chlamydomonas reinhardtii requires a lipase with broad substrate specificity. Eukaryotic cell, 11(12), 1451-1462.

Li, X., Moellering, E. R., Liu, B., Johnny, C., Fedewa, M., Sears, B. B. ... & Benning, C. (2012). A galactoglycerolipid lipase is required for triacylglycerol accumulation and survival following nitrogen deprivation in Chlamydomonas reinhardtii. The Plant Cell, 24(11), 4670-4686.

Lin, W., Zhao, D. & Luo, J. (2018). Distribution of alkaline phosphatase genes in cyanobacteria and the role of alkaline phosphatase on the acquisition of phosphorus from dissolved organic phosphorus for cyanobacterial growth. Journal of Applied Phycology, 30(2), 839-850.

Liu, S. H., Shen, C. C., Yi, Y. C., Tsai, J. J., Wang, C. C., Chueh, J. T., ... & Sheu, M. L. (2010). Honokiol inhibits gastric tumourigenesis by activation of 15-lipoxygenase-1 and consequent inhibition of peroxisome proliferator-activated receptor-γ and COX-2-dependent signals. British journal of pharmacology, 160(8), 1963-1972.

Lockau, W., Massalsky, B. & Dirmeier, A. (1988). Purification and partial characterization of a calcium-stimulated protease from the cyanobacterium, Anabaena variabilis. European journal of biochemistry, 172(2), 433-438.

Lolans, K., Calvert, K., Won, S., Clark, J. & Hayden, M. K. (2010). Direct ertapenem disk screening method for identification of KPC-producing Klebsiella pneumoniae and Escherichia coli in surveillance swab specimens. Journal of clinical microbiology, 48(3), 836-841.

Lu, X., Zhang, J., Liu, S., Zhang, D., Xu, Z., Wu, J. ... & Chen, J. (2013). Overproduction, purification, and characterization of extracellular lipoxygenase of Pseudomonas aeruginosa in Escherichia coli. Applied microbiology and biotechnology, 97(13), 5793-5800.

Lubitz, W., Reijerse, E. & van Gastel, M. (2007). [NiFe] and [FeFe] hydrogenases studied by advanced magnetic resonance techniques. Chemical reviews, 107(10), 4331-4365.

Luo, D. & Rando, T. A. (2003). The regulation of catalase gene expression in mouse muscle cells is dependent on the CCAAT-binding factor NF-Y. Biochemical and biophysical research communications, 303(2), 609-618.

Ma, J., Wang, P., Ren, L., Wang, X. & Paerl, H. W. (2019). Using alkaline phosphatase activity as a supplemental index to optimize predicting algal blooms in phosphorus-deficient lakes: A case study of Lake Taihu, China. Ecological Indicators, 103, 698-712.

Martínková, L., Rucká, L., Nešvera, J. & Pátek, M. (2017). Recent advances and challenges in the heterologous production of microbial nitrilases for biocatalytic applications. World Journal of Microbiology and Biotechnology, 33(1), 1-11.

Mishra, A. K., Kaushik, M. S. & Tiwari, D. N. (2019). Nitrogenase and hydrogenase: enzymes for nitrogen fixation and hydrogen production in cyanobacteria. In Cyanobacteria (pp. 173-191). Academic Press.

Mogharabi, M. & Faramarzi, M. A. (2016). Are algae the future source of enzymes? Trends in Peptide and Protein Sciences, 1(1), 1-6.

Mohsenpour, S. F., Hennige, S., Willoughby, N., Adeloye, A. & Gutierrez, T. (2020). Integrating micro-algae into wastewater treatment: A review. Science of The Total Environment, 142168.

Moroney, J. V., Bartlett, S. G. & Samuelsson, G. (2001). Carbonic anhydrases in plants and algae. Plant, Cell & Environment, 24(2), 141-153.

Ng, I. S., Chen, T., Lin, R., Zhang, X., Ni, C. & Sun, D. (2014). Decolorization of textile azo dye and Congo red by an isolated strain of the dissimilatory manganese-reducing bacterium Shewanella xiamenensis BC01. Applied microbiology and biotechnology, 98(5), 2297-2308.

Okuno, Y., Matsuda, M., Kobayashi, H., Morita, K., Suzuki, E., Fukuhara, A. ... & Shimomura, I. (2008). Adipose expression of catalase is regulated via a novel remote PPARγ-responsive region. Biochemical and biophysical research communications, 366(3), 698-704.

Patel, D. D., Patel, R. R. & Thakkar, V. R. (2015). Purification, characterization and application of lipoxygenase isoenzymes from Lasiodiplodia theobromae. Applied biochemistry and biotechnology, 175(1), 513-525.

Patel, V. R. & Bhatt, N. (2020). Application of stress induces ascorbate peroxidases of S. polyrhiza for green-synthesis Cu nanoparticles. Arabian Journal of Chemistry, 13(12), 8783-8792.

Pervez, S., Shahid, F., Aman, A. & Qader, S. A. U. (2017). Algal biomass: a sustainable, economical and renewable approach for microbial production of pectinolytic enzymes using submerged and solid-state fermentation techniques. Biocatalysis and Biotransformation, 35(6), 442-449.

Peters, J. W., Lanzilotta, W. N., Lemon, B. J. & Seefeldt, L. C. (1998). X-ray crystal structure of the Fe-only hydrogenase (CpI) from Clostridium pasteurianum to 1.8 angstrom resolution. Science, 282(5395), 1853-1858.

Piotrowski, M. (2008). Primary or secondary? Versatile nitrilases in plant metabolism. Phytochemistry, 69(15), 2655-2667.

Pirinççi, N., Kaya, T. Y., Kaba, M., Ozan, T., Geçit, İ., Özveren, H. ... & Ceylan, K. (2017). Serum adenosine deaminase, catalase, and carbonic anhydrase activities in patients with renal cell carcinoma. Redox Report, 22(6), 252-256.

Polle, J. E., Tran, D. & Ben-Amotz, A. (2009). History, distribution, and habitats of algae of the genus Dunaliella Teodoresco (Chlorophyceae). The alga Dunaliella, 1-14.

Pragya, N., Pandey, K. K., & Sahoo, P. K. (2013). A review on harvesting, oil extraction and biofuels production technologies from microalgae. Renewable and sustainable energy reviews, 24, 159-171.

Price, S., Kuzhiumparambil, U., Pernice, M. & Ralph, P. J. (2020). Cyanobacterial polyhydroxybutyrate for sustainable bioplastic production: Critical review and perspectives. Journal of Environmental Chemical Engineering, 8(4), 104007.

Qeshmi, F. I., Homaei, A., Fernandes, P. & Javadpour, S. (2018). Marine microbial L-asparaginase: biochemistry, molecular approaches and applications in tumor therapy and in food industry. Microbiological research, 208, 99-112.

Radakovits, R., Jinkerson, R. E., Darzins, A. & Posewitz, M. C. (2010). Genetic engineering of algae for enhanced biofuel production. Eukaryotic cell, 9(4), 486-501.

Ratnam, D. V., Ankola, D. D., Bhardwaj, V., Sahana, D. K. & Kumar, M. R. (2006). Role of antioxidants in prophylaxis and therapy: A pharmaceutical perspective. Journal of controlled release, 113(3), 189-207.

Rezayian, M., Niknam, V. & Ebrahimzadeh, H. (2019). Oxidative damage and antioxidative system in algae. Toxicology reports, 6, 1309-1313.

Rodriguez, R. J., White Jr, J. F., Arnold, A. E. & Redman, A. R. A. (2009). Fungal endophytes: diversity and functional roles. New phytologist, 182(2), 314-330.

Rosenberg, Z. M. M. (2006). Current trends of β-galactosidase application in food technology. J Food Nutr Res, 45(2), 47-54.

Sahoo, R., Kumar, S. & Ahuja, P. S. (2001). Induction of a new isozyme of superoxide dismutase at low temperature in Potentilla astrisanguinea Lodd. variety argyrophylla (Wall. ex. Lehm) Griers. Journal of plant physiology, 158(8), 1093-1097.

Saini, D. K., Chakdar, H., Pabbi, S. & Shukla, P. (2020). Enhancing production of microalgal biopigments through metabolic and genetic engineering. Critical reviews in food science and nutrition, 60(3), 391-405.

Sakthinathan, B. (2018). Economic Benefits of Aquatic Plants - A Review. Mod App Ocean & etr Sci 1(3). MAOPS.MS.ID.000115.

Shechter, D. (2016). French Press Cell Lysis: How does it compare to Homogenization. https://www.beei.com/blog/french-press-cell-lysis-how-does-it-compare-to homogenization

Shen, J. D., Cai, X., Liu, Z. Q. & Zheng, Y. G. (2021). Nitrilase: a promising biocatalyst in industrial applications for green chemistry. Critical Reviews in Biotechnology, 41(1), 72-93.

Simair, A. A., Khushk, I., Qureshi, A. S., Bhutto, M. A., Chaudhry, H. A., Ansari, K. A. & Lu, C. (2017). Amylase production from thermophilic Bacillus sp. BCC 021-50 isolated from a marine environment. Fermentation, 3(2), 25.

Singh, R. S., Sing, T. & Pandey, A. (2019). Microbial enzyme-an over view:Advances in enzyme technology (chapter-1). Elsevier.

Singh, S. P. & Singh, P. (2015). Effect of temperature and light on the growth of algae species: a review. Renewable and sustainable energy reviews, 50, 431-444.

Souza, P. M. D. & Magalhães, P. D. O. (2010). Application of microbial α-amylase in industry-A review. Brazilian journal of microbiology, 41, 850-861.

Spier, M. R., Peron-Schlosser, B., Paludo, L. C., Gallo-García, L. A. & Zanette, C. M. (2020). Microalgae as enzymes biofactories. In Handbook of Microalgae-Based Processes and Products (pp. 687-706). Academic Press.

Subhadra, B. G. & Edwards, M. (2011). Coproduct market analysis and water footprint of simulated commercial algal biorefineries. Applied Energy, 88(10), 3515-3523.

Sunder, A. V., Shah, S., Rayavarapu, P. & Wangikar, P. P. (2020). Expanding the repertoire of nitrilases with broad substrate specificity and high substrate tolerance for biocatalytic applications. Process Biochemistry, 94, 289-296.

Suwal, S., Bentahar, J., Marciniak, A., Beaulieu, L., Deschênes, J. S. & Doyen, A. (2019). Evidence of the production of galactooligosaccharide from whey permeate by the microalgae Tetradesmus obliquus. Algal Research, 39, 101470.

Swarnalatha, G. V., Hegde, N. S., Chauhan, V. S. & Sarada, R. (2015). The effect of carbon dioxide rich environment on carbonic anhydrase activity, growth and metabolite production in indigenous freshwater microalgae. Algal research, 9, 151-159.

Tekaya, N., Saiapina, O., Ouada, H. B., Lagarde, F., Ouada, H. B. & Jaffrezic-Renault, N. (2013). Ultra-sensitive conductometric detection of heavy metals based on inhibition of alkaline phosphatase activity from Arthrospira platensis. Bioelectrochemistry, 90, 24-29.

Tian, L. (2015). Carotenoids, genetically modified foods, and vitamin A nutrition. In Genetically Modified Organisms in Food: Production, Safety, Regulation and Public Health (pp. 353-360). Elsevier Inc..

Tokuoka, M., Sawamura, N., Kobayashi, K. & Mizuno, A. (2010). Simple metabolite extraction method for metabolic profiling of the solid-state fermentation of Aspergillus oryzae. Journal of bioscience and bioengineering, 110(6), 665-669.

Trivedi, J., Aila, M., Bangwal, D. P., Kaul, S. & Garg, M. O. (2015). Algae based biorefinery—how to make sense? Renewable and Sustainable Energy Reviews, 47, 295-307

Ubando, A. T., Culaba, A. B., Tan, R. R. & Ng, D. K. (2012). A systematic approach for optimization of an algal biorefinery using fuzzy linear programming. In Computer Aided Chemical Engineering (Vol. 31, pp. 805-809). Elsevier.

Uggetti, E., Sialve, B., Hamelin, J., Bonnafous, A. & Steyer, J. P. (2018). CO2 addition to increase biomass production and control microalgae species in high-rate algal ponds treating wastewater. Journal of CO2 Utilization, 28, 292-298.

Varjani, S., Pandey, A., Gnansounou, E., Khanal, S. K. & Raveendran, S. (Eds.). (2020). Current Developments in Biotechnology and Bioengineering: Resource Recovery from Wastes. Elsevier.

Vingiani, G. M., De Luca, P., Ianora, A., Dobson, A. D. & Lauritano, C. (2019). Microalgal enzymes with biotechnological applications. Marine drugs, 17(8), 459.

Wang, H. M. D., Chen, C. C., Huynh, P. & Chang, J. S. (2015). Exploring the potential of using algae in cosmetics. Bioresource technology, 184, 355-362.

Weaver, P. F., Lien, S. & Seibert, M. (1980). Photobiological production of hydrogen. Solar Energy, 24(1), 3-45.

Weignerová, L., Simerská, P. & Křen, V. (2009). α-Galactosidases and their applications in biotransformations. Biocatalysis and Biotransformation, 27(2), 79-89.

Wells, M. L., Potin, P., Craigie, J. S., Raven, J. A., Merchant, S. S., Helliwell, K. E. ... & Brawley, S. H. (2017). Algae as nutritional and functional food sources: revisiting our understanding. Journal of applied phycology, 29(2), 949-982.

Xie, C., Li, B., Xu, Y., Ji, D. & Chen, C. (2013). Characterization of the global transcriptome for Pyropia haitanensis (Bangiales, Rhodophyta) and development of cSSR markers. BMC genomics, 14(1), 1-13.

Yang, Y., Shi, X., Ballent, W. & Mayer, B. K. (2017). Biological Phosphorus Recovery: Review of Current Progress and Future Needs: Yang et al. Water Environment Research, 89(12), 2122-2135.

Yen, H. W., Chen, P. W., Hsu, C. Y. & Lee, L. (2017). The use of autotrophic Chlorella vulgaris in chromium (VI) reduction under different reduction conditions. Journal of the Taiwan Institute of Chemical Engineers, 74, 1-6.

Yoboue, E. D., Mougeolle, A., Kaiser, L., Averet, N., Rigoulet, M. & Devin, A. (2014). The role of mitochondrial biogenesis and ROS in the control of energy supply in proliferating cells. Biochimica et Biophysica Acta (BBA)-Bioenergetics, 1837(7), 1093-1098.

Yu, S. & Pedersen, M. (1990). The effect of salinity changes on the activity of α-galactosidase of the red algae Gracilaria sordida and G. tenuistipitata.

Zanette, C. M., Mariano, A. B., Yukawa, Y. S., Mendes, I. & Spier, M. R. (2019). Microalgae mixotrophic cultivation for β-galactosidase production. Journal of Applied Phycology, 31(3), 1597-1606.

Zelko, I. N., Mariani, T. J. & Folz, R. J. (2002). Superoxide dismutase multigene family: a comparison of the CuZn-SOD (SOD1), Mn-SOD (SOD2), and EC-SOD (SOD3) gene structures, evolution, and expression. Free Radical Biology and Medicine, 33(3), 337-349.

Zhang, L., Yin, B., Wang, C., Jiang, S., Wang, H., Yuan, Y. A. & Wei, D. (2014). Structural insights into enzymatic activity and substrate specificity determination by a single amino acid in nitrilase from Synechocystis sp. PCC6803. Journal of structural biology, 188(2), 93-101.

Zhang, Y., Huang, Z., Zheng, H., Wang, Q. & Li, A. (2020). Growth, biochemical composition and photosynthetic performance of Scenedesmus acuminatus under different initial sulfur supplies. Algal Research, 45, 101728.

INDEX

C

D

E

F

S

T

U

V

W

X

Y